THE CELEBRATED JUMPING FROG OF CALAVERAS COUNTY,

and Other Sketches

THE OXFORD MARK TWAIN

Shelley Fisher Fishkin, Editor

The Celebrated Jumping Frog of Calaveras County, and Other Sketches
Introduction: Roy Blount Jr.
Afterword: Richard Bucci

The Innocents Abroad
Introduction: Mordecai Richler
Afterword: David E. E. Sloane

Roughing It
Introduction: George Plimpton
Afterword: Henry B. Wonham

The Gilded Age
Introduction: Ward Just
Afterword: Gregg Camfield

Sketches, New and Old
Introduction: Lee Smith
Afterword: Sherwood Cummings

The Adventures of Tom Sawyer
Introduction: E. L. Doctorow
Afterword: Albert E. Stone

A Tramp Abroad
Introduction: Russell Banks
Afterword: James S. Leonard

The Prince and the Pauper

Introduction: Judith Martin

Afterword: Everett Emerson

Life on the Mississippi

Introduction: Willie Morris

Afterword: Lawrence Howe

Adventures of Huckleberry Finn

Introduction: Toni Morrison

Afterword: Victor A. Doyno

A Connecticut Yankee in King Arthur's Court

Introduction: Kurt Vonnegut, Jr.

Afterword: Louis J. Budd

Merry Tales

Introduction: Anne Bernays

Afterword: Forrest G. Robinson

The American Claimant

Introduction: Bobbie Ann Mason

Afterword: Peter Messent

The £1,000,000 Bank-Note and Other New Stories

Introduction: Malcolm Bradbury

Afterword: James D. Wilson

Tom Sawyer Abroad

Introduction: Nat Hentoff

Afterword: M. Thomas Inge

The Tragedy of Pudd'nhead Wilson and the Comedy Those Extraordinary Twins

Introduction: Sherley Anne Williams

Afterword: David Lionel Smith

Personal Recollections of Joan of Arc
Introduction: Justin Kaplan
Afterword: Susan K. Harris

The Stolen White Elephant and Other Detective Stories
Introduction: Walter Mosley
Afterword: Lillian S. Robinson

How to Tell a Story and Other Essays
Introduction: David Bradley
Afterword: Pascal Covici, Jr.

Following the Equator and Anti-imperialist Essays
Introduction: Gore Vidal
Afterword: Fred Kaplan

The Man That Corrupted Hadleyburg and Other Stories and Essays
Introduction: Cynthia Ozick
Afterword: Jeffrey Rubin-Dorsky

The Diaries of Adam and Eve
Introduction: Ursula K. Le Guin
Afterword: Laura E. Skandera-Trombley

What Is Man?
Introduction: Charles Johnson
Afterword: Linda Wagner-Martin

The $30,000 Bequest and Other Stories
Introduction: Frederick Busch
Afterword: Judith Yaross Lee

Christian Science
Introduction: Garry Wills
Afterword: Hamlin Hill

Chapters from My Autobiography
Introduction: Arthur Miller
Afterword: Michael J. Kiskis

1601, and Is Shakespeare Dead?
Introduction: Erica Jong
Afterword: Leslie A. Fiedler

Extract from Captain Stormfield's Visit to Heaven
Introduction: Frederik Pohl
Afterword: James A. Miller

Speeches
Introduction: Hal Holbrook
Afterword: David Barrow

The Celebrated Jumping Frog of Calaveras County, and Other Sketches

Mark Twain

FOREWORD

SHELLEY FISHER FISHKIN

INTRODUCTION

ROY BLOUNT JR.

AFTERWORD

RICHARD BUCCI

New York Oxford

OXFORD UNIVERSITY PRESS

1996

OXFORD UNIVERSITY PRESS

Oxford New York

Athens, Auckland, Bangkok, Bogotá, Bombay
Buenos Aires, Calcutta, Cape Town, Dar es Salaam
Delhi, Florence, Hong Kong, Istanbul, Karachi
Kuala Lumpur, Madras, Madrid, Melbourne
Mexico City, Nairobi, Paris, Singapore
Taipei, Tokyo, Toronto
and associated companies in
Berlin, Ibadan

Text design by Richard Hendel
Composition: David Thorne

Published by
Oxford University Press, Inc.
198 Madison Avenue, New York,
New York 10016

Library of Congress
Cataloging-in-Publication Data

Twain, Mark, 1835–1910.
The celebrated jumping frog of Calaveras County
and other sketches / Mark Twain;
introduction: Roy Blount, Jr.,
afterword: Richard Bucci.
p. cm. — (The Oxford Mark Twain)
1. Humorous stories, American. I. Title. II. Series:
Twain, Mark, 1835–1910. Works. 1996.
PS1322.C4 1996
813'. 4—dc20
96—12305
CIP
ISBN 0-19-510131-6 (trade ed.)
ISBN 0-19-511400-0 (lib. ed.)
ISBN 0-19-509088-8 (trade ed. set)
ISBN 0-19-511345-4 (lib. ed. set)

9 8 7 6 5 4 3 2 1

Printed in the United States of America
on acid-free paper

FRONTISPIECE

This photograph of Samuel L. Clemens is from a *carte de visite* that was taken at Edouart's Photographic Gallery in San Francisco around 1865, when Clemens wrote and first published the "Jumping Frog" story. (Courtesy, The Mark Twain Papers, The Bancroft Library)

CONTENTS

EDITOR'S NOTE

The Oxford Mark Twain consists of twenty-nine volumes of facsimiles of the first American editions of Mark Twain's works, with an editor's foreword, new introductions, afterwords, notes on the texts, and essays on the illustrations in volumes with artwork. The facsimiles have been reproduced from the originals unaltered, except that blank pages in the front and back of the books have been omitted, and any seriously damaged or missing pages have been replaced by pages from other first editions (as indicated in the notes on the texts).

In the foreword, introduction, afterword, and essays on the illustrations, the titles of Mark Twain's works have been capitalized according to modern conventions, as have the names of characters (except where otherwise indicated). In the case of discrepancies between the title of a short story, essay, or sketch as it appears in the original table of contents and as it appears on its own title page, the title page has been followed. The parenthetical numbers in the introduction, afterwords, and illustration essays are page references to the facsimiles.

FOREWORD

Shelley Fisher Fishkin

Samuel Clemens entered the world and left it with Halley's Comet, little dreaming that generations hence Halley's Comet would be less famous than Mark Twain. He has been called the American Cervantes, our Homer, our Tolstoy, our Shakespeare, our Rabelais. Ernest Hemingway maintained that "all modern American literature comes from one book by Mark Twain called *Huckleberry Finn*." President Franklin Delano Roosevelt got the phrase "New Deal" from *A Connecticut Yankee in King Arthur's Court. The Gilded Age* gave an entire era its name. "The future historian of America," wrote George Bernard Shaw to Samuel Clemens, "will find your works as indispensable to him as a French historian finds the political tracts of Voltaire."[1]

There is a Mark Twain Bank in St. Louis, a Mark Twain Diner in Jackson Heights, New York, a Mark Twain Smoke Shop in Lakeland, Florida. There are Mark Twain Elementary Schools in Albuquerque, Dayton, Seattle, and Sioux Falls. Mark Twain's image peers at us from advertisements for Bass Ale (his drink of choice was Scotch), for a gas company in Tennessee, a hotel in the nation's capital, a cemetery in California.

Ubiquitous though his name and image may be, Mark Twain is in no danger of becoming a petrified icon. On the contrary: Mark Twain lives. *Huckleberry Finn* is "the most taught novel, most taught long work, and most taught piece of American literature" in American schools from junior high to the graduate level.[2] Hundreds of Twain impersonators appear in theaters, trade shows, and shopping centers in every region of the country.[3] Scholars publish hundreds of articles as well as books about Twain every year, and he is the subject of daily exchanges on the Internet. A journalist somewhere in the world finds a reason to quote Twain just about every day. Television series such as *Bonanza*, *Star Trek: The Next Generation*, and *Cheers* broadcast episodes that feature Mark Twain as a character. Hollywood screenwriters regularly produce movies inspired by his works, and writers of mysteries and science fiction continue to weave him into their plots.[4]

A century after the American Revolution sent shock waves throughout Europe, it took Mark Twain to explain to Europeans and to his countrymen alike what that revolution had wrought. He probed the significance of this new land and its new citizens, and identified what it was in the Old World that America abolished and rejected. The founding fathers had thought through the political dimensions of making a new society; Mark Twain took on the challenge of interpreting the social and cultural life of the United States for those outside its borders as well as for those who were living the changes he discerned.

Americans may have constructed a new society in the eighteenth century, but they articulated what they had done in voices that were largely interchangeable with those of Englishmen until well into the nineteenth century. Mark Twain became the voice of the new land, the leading translator of what and who the "American" was — and, to a large extent, is. Frances Trollope's *Domestic Manners of the Americans,* a best-seller in England, Hector St. John de Crèvecoeur's *Letters from an American Farmer,* and Tocqueville's *Democracy in America* all tried to explain America to Europeans. But Twain did more than that: he allowed European readers to *experience* this strange "new world." And he gave his countrymen the tools to do two things they had not quite had the confidence to do before. He helped them stand before the cultural icons of the Old World unembarrassed, unashamed of America's lack of palaces and shrines, proud of its brash practicality and bold inventiveness, unafraid to reject European models of "civilization" as tainted or corrupt. And he also helped them recognize their own insularity, boorishness, arrogance, or ignorance, and laugh at it — the first step toward transcending it and becoming more "civilized," in the best European sense of the word.

Twain often strikes us as more a creature of our time than of his. He appreciated the importance and the complexity of mass tourism and public relations, fields that would come into their own in the twentieth century but were only fledgling enterprises in the nineteenth. He explored the liberating potential of humor and the dynamics of friendship, parenting, and marriage. He narrowed the gap between "popular" and "high" culture, and he meditated on the enigmas of personal and national identity. Indeed, it would be difficult to find an issue on the horizon today that Twain did not touch on somewhere in his work. Heredity versus environment? Animal rights? The boundaries of gender? The place of black voices in the cultural heritage of the United States? Twain was there.

With startling prescience and characteristic grace and wit, he zeroed in on many of the key challenges — political, social, and technological — that would face his country and the world for the next hundred years: the challenge of race relations in a society founded on both chattel slavery and ideals of equality, and the intractable problem of racism in American life; the potential of new technologies to transform our lives in ways that can be both exhilarating and terrifying — as well as unpredictable; the problem of imperialism and the difficulties entailed in getting rid of it. But he never lost sight of the most basic challenge of all: each man or woman's struggle for integrity in the face of the seductions of power, status, and material things.

Mark Twain's unerring sense of the right word and not its second cousin taught people to pay attention when he spoke, in person or in print. He said things that were smart and things that were wise, and he said them incomparably well. He defined the rhythms of our prose and the contours of our moral map. He saw our best and our worst, our extravagant promise and our stunning failures, our comic foibles and our tragic flaws. Throughout the world he is viewed as the most distinctively American of American authors — and as one of the most universal. He is assigned in classrooms in Naples, Riyadh, Belfast, and Beijing, and has been a major influence on twentieth-century writers from Argentina to Nigeria to Japan. The Oxford Mark Twain celebrates the versatility and vitality of this remarkable writer.

The Oxford Mark Twain reproduces the first American editions of Mark Twain's books published during his lifetime.[5] By encountering Twain's works in their original format—typography, layout, order of contents, and illustrations—readers today can come a few steps closer to the literary artifacts that entranced and excited readers when the books first appeared. Twain approved of and to a greater or lesser degree supervised the publication of all of this material.[6] The Mark Twain House in Hartford, Connecticut, generously loaned us its originals.[7] When more than one copy of a first American edition was available, Robert H. Hirst, general editor of the Mark Twain Project, in cooperation with Marianne Curling, curator of the Mark Twain House (and Jeffrey Kaimowitz, head of Rare Books for the Watkinson Library of Trinity College, Hartford, where the Mark Twain House collection is kept), guided our decision about which one to use.[8] As a set, the volumes also contain more than eighty essays commissioned especially for The Oxford Mark Twain, in which distinguished contributors reassess Twain's achievement as a writer and his place in the cultural conversation that he did so much to shape.

Each volume of The Oxford Mark Twain is introduced by a leading American, Canadian, or British writer who responds to Twain—often in a very personal way—as a fellow writer. Novelists, journalists, humorists, columnists, fabulists, poets, playwrights—these writers tell us what Twain taught them and what in his work continues to speak to them. Reading Twain's books, both famous and obscure, they reflect on the genesis of his art and the characteristics of his style, the themes he illuminated, and the aesthetic strategies he pioneered. Individually and collectively their contributions testify to the place Mark Twain holds in the hearts of readers of all kinds and temperaments.

Scholars whose work has shaped our view of Twain in the academy today have written afterwords to each volume, with suggestions for further reading. Their essays give us a sense of what was going on in Twain's life when he wrote the book at hand, and of how that book fits into his career. They explore how each book reflects and refracts contemporary events, and they show Twain responding to literary and social currents of the day, variously accepting, amplifying, modifying, and challenging prevailing paradigms. Sometimes they argue that works previously dismissed as quirky or eccentric departures actually address themes at the heart of Twain's work from the start. And as they bring new perspectives to Twain's composition strategies in familiar texts, several scholars see experiments in form where others saw only formlessness, method where prior critics saw only madness. In addition to elucidating the work's historical and cultural context, the afterwords provide an overview of responses to each book from its first appearance to the present.

Most of Mark Twain's books involved more than Mark Twain's words: unique illustrations. The parodic visual send-ups of "high culture" that Twain himself drew for *A Tramp Abroad*, the sketch of financial manipulator Jay Gould as a greedy and sadistic "Slave Driver" in *A Connecticut Yankee in King Arthur's Court*, and the memorable drawings of Eve in *Eve's Diary* all helped Twain's books to be sold, read, discussed, and preserved. In their essays for each volume that contains artwork, Beverly R. David and Ray Sapirstein highlight the significance of the sketches, engravings, and photographs in the first American editions of Mark Twain's works, and tell us what is known about the public response to them.

The Oxford Mark Twain invites us to read some relatively neglected works by Twain in the company of some of the most engaging literary figures of our time. Roy Blount Jr., for example, riffs in a deliciously Twain-like manner on "An Item Which the Editor Himself Could Not Understand," which may well rank as one of the least-known pieces Twain ever published. Bobbie Ann Mason celebrates the "mad energy" of Twain's most obscure comic novel, *The American Claimant*, in which the humor "hurtles beyond tall tale into simon-pure absurdity."[9] Garry Wills finds that *Christian Science* "gets us very close to the heart of American culture." Lee Smith reads "Political Economy" as a sharp and funny essay on language. Walter Mosley sees "The Stolen White Elephant," a story "reduced to a series of ridiculous telegrams related by an untrustworthy narrator caught up in an adventure that is as impossible as it is ludicrous," as a stunningly compact and economical satire of a world we still recognize as our own. Anne Bernays returns to "The Private History of a Campaign That Failed" and finds "an antiwar manifesto that is also confession, dramatic monologue, a plea for understanding and absolution, and a romp that gradually turns into atrocity even as we watch." After revisiting Captain Stormfield's heaven, Frederik Pohl finds that there "is no imaginable place more pleasant to spend eternity." Indeed, Pohl writes, "one would almost be willing to die to enter it."

While less familiar works receive fresh attention in The Oxford Mark Twain, new light is cast on the best-known works as well. Judith Martin ("Miss Manners") points out that it is by reading a court etiquette book that Twain's pauper learns how to behave as a proper prince. As important as etiquette may be in the palace, Martin notes, it is even more important in the slums.

> That etiquette is a sorer point with the ruffians in the street than with the proud dignitaries of the prince's court may surprise some readers. As in our own streets, etiquette is always a more volatile subject among those who cannot count on being treated with respect than among those who have the power to command deference.

And taking a fresh look at *Adventures of Huckleberry Finn,* Toni Morrison writes,

> much of the novel's genius lies in its quiescence, the silences that pervade it and give it a porous quality that is by turns brooding and soothing. It lies in . . . the subdued images in which the repetition of a simple word, such as "lonesome," tolls like an evening bell; the moments when nothing is said, when scenes and incidents swell the heart unbearably precisely because unarticulated, and force an act of imagination almost against the will.

Engaging Mark Twain as one writer to another, several contributors to The Oxford Mark Twain offer new insights into the processes by which his books came to be. Russell Banks, for example, reads *A Tramp Abroad* as "an important revision of Twain's incomplete first draft of *Huckleberry Finn*, a second draft, if you will, which in turn made possible the third and final draft." Erica Jong suggests that *1601,* a freewheeling parody of Elizabethan manners and mores, written during the same summer Twain began *Huckleberry Finn*, served as "a warm-up for his creative process" and "primed the pump for other sorts of freedom of expression." And Justin Kaplan suggests that "one of the transcendent figures standing behind and shaping" *Joan of Arc* was Ulysses S. Grant, whose memoirs Twain had recently published, and who, like Joan, had risen unpredictably "from humble and obscure origins" to become a "military genius" endowed with "the gift of command, a natural eloquence, and an equally natural reserve."

As a number of contributors note, Twain was a man ahead of his times. *The Gilded Age* was the first "Washington novel," Ward Just tells us, because "Twain was the first to see the possibilities that had eluded so many others." Commenting on *The Tragedy of Pudd'nhead Wilson,* Sherley Anne Williams observes that "Twain's argument about the power of environment in shaping character runs directly counter to prevailing sentiment where the negro was concerned." Twain's fictional technology, wildly fanciful by the standards of his day, predicts developments we take for granted in ours. DNA cloning, fax machines, and photocopiers are all prefigured, Bobbie Ann Mason tells us, in *The American Claimant.* Cynthia Ozick points out that the "telelectrophonoscope" we meet in "From the 'London Times' of 1904" is suspiciously like what we know as "television." And Malcolm Bradbury suggests that in the "phrenophones" of "Mental Telegraphy" "the Internet was born."

Twain turns out to have been remarkably prescient about political affairs as well. Kurt Vonnegut sees in *A Connecticut Yankee* a chilling foreshadowing (or perhaps a projection from the Civil War) of "all the high-tech atrocities which followed, and which follow still." Cynthia Ozick suggests that "The Man That Corrupted Hadleyburg," along with some of the other pieces collected under that title — many of them written when Twain lived in a Vienna ruled by Karl Lueger, a demagogue Adolf Hitler would later idolize — shoot up moral flares that shed an eerie light on the insidious corruption, prejudice, and hatred that reached bitter fruition under the Third Reich. And Twain's portrait in this book of "the dissolving Austria-Hungary of the 1890s," in Ozick's view, presages not only the Sarajevo that would erupt in 1914 but also "the disintegrated components of the former Yugoslavia" and "the *fin-de-siècle* Sarajevo of our own moment."

Despite their admiration for Twain's ambitious reach and scope, contributors to The Oxford Mark Twain also recognize his limitations. Mordecai Richler, for example, thinks that "the early pages of *Innocents Abroad* suffer from being a tad broad, proffering more burlesque than inspired satire," perhaps because Twain was "trying too hard for knee-slappers." Charles Johnson notes that the Young Man in Twain's philosophical dialogue about free will and determinism (*What Is Man?*) "caves in far too soon," failing to challenge what through late-twentieth-century eyes looks like "pseudoscience" and suspect essentialism in the Old Man's arguments.

Some contributors revisit their first encounters with Twain's works, recalling what surprised or intrigued them. When David Bradley came across "Fenimore Cooper's Literary Offences" in his college library, he "did not at first realize that Twain was being his usual ironic self with all this business about the 'nineteen rules governing literary art in the domain of romantic fiction,' but by the time I figured out there was no such list outside Twain's own head, I had decided that the rules made *sense*. . . . It seemed to me they were a pretty good blueprint for writing — Negro writing included." Sherley Anne Williams remembers that part of what attracted her to *Pudd'nhead Wilson* when she first read it thirty years ago was "that Twain, writing at the end of the nineteenth century, could imagine negroes as characters, albeit white ones, who actually thought for and of themselves, whose actions were the product of their thinking rather than the spontaneous ephemera of physical instincts that stereotype assigned to blacks." Frederik Pohl recalls his first reading of *Huckleberry Finn* as "a watershed event" in his life, the first book he read as a child in which "bad people" ceased to exercise a monopoly on doing "bad things." In *Huckleberry Finn* "some seriously bad things — things like the possession and mistreatment of black slaves, like stealing and lying, even like killing other people in duels — were quite often done by people who not only thought of themselves as exemplarily moral but, by any other standards I knew how to apply, actually *were* admirable citizens." The world that

Tom and Huck lived in, Pohl writes, "was filled with complexities and contradictions," and resembled "the world I appeared to be living in myself."

Other contributors explore their more recent encounters with Twain, explaining why they have revised their initial responses to his work. For Toni Morrison, parts of *Huckleberry Finn* that she "once took to be deliberate evasions, stumbles even, or a writer's impatience with his or her material," now strike her "as otherwise: as entrances, crevices, gaps, seductive invitations flashing the possibility of meaning. Unarticulated eddies that encourage diving into the novel's undertow — the real place where writer captures reader." One such "eddy" is the imprisonment of Jim on the Phelps farm. Instead of dismissing this portion of the book as authorial bungling, as she once did, Morrison now reads it as Twain's commentary on the 1880s, a period that "saw the collapse of civil rights for blacks," a time when "the nation, as well as Tom Sawyer, was deferring Jim's freedom in agonizing play." Morrison believes that Americans in the 1880s were attempting "to bury the combustible issues Twain raised in his novel," and that those who try to kick Huck Finn out of school in the 1990s are doing the same: "The cyclical attempts to remove the novel from classrooms extend Jim's captivity on into each generation of readers."

Although imitation-Hemingway and imitation-Faulkner writing contests draw hundreds of entries annually, no one has ever tried to mount a faux-Twain competition. Why? Perhaps because Mark Twain's voice is too much a part of who we are and how we speak even today. Roy Blount Jr. suggests that it is impossible, "at least for an American writer, to parody Mark Twain. It would be like doing an impression of your father or mother: he or she is already there in your voice."

Twain's style is examined and celebrated in The Oxford Mark Twain by fellow writers who themselves have struggled with the nuances of words, the structure of sentences, the subtleties of point of view, and the trickiness of opening lines. Bobbie Ann Mason observes, for example, that "Twain loved the sound of words and he knew how to string them by sound, like different shades of one color: 'The earl's barbaric eye,' 'the Usurping Earl,' 'a double-dyed humbug.'" Twain "relied on the punch of plain words" to show writers how to move beyond the "wordy romantic rubbish" so prevalent in nineteenth-century fiction, Mason says; he "was one of the first writers in America to deflower literary language." Lee Smith believes that "American writers have benefited as much from the way Mark Twain opened up the possibilities of first-person narration as we have from his use of vernacular language." (She feels that "the ghost of Mark Twain was hovering someplace in the background" when she decided to write her novel *Oral History* from the standpoint of multiple first-person narrators.) Frederick Busch maintains that "A Dog's Tale" "boasts one of the great opening sentences" of all time: "My father was a St. Bernard, my mother was a collie, but I am a Presbyterian." And Ursula Le Guin marvels at the ingenuity of the following sentence that she encounters in *Extracts from Adam's Diary*.

> . . . This made her sorry for the creatures which live in there, which she calls fish, for she continues to fasten names on to things that don't need them and don't come when they are called by them, which is a matter of no consequence to her, as she is such a numskull anyway; so she got a lot of them out and brought them in last night and put them in my bed to keep warm, but I have noticed them now and then all day, and I don't see that they are any happier there than they were before, only quieter.[10]

Le Guin responds,

> Now, that is a pure Mark-Twain-tour-de-force sentence, covering an immense amount of territory in an effortless, aimless ramble that seems to be heading nowhere in particular and ends up with breathtaking accuracy at the gold mine. Any sensible child would find that funny, perhaps not following all its divagations but delighted by the swing of it, by the word "numskull," by the idea of putting fish in the bed; and as that child grew older and reread it, its reward would only grow; and if that grown-up child had to write an essay on the piece and therefore earnestly studied and pored over this sentence, she would end up in unmitigated admiration of its vocabulary, syntax, pacing, sense, and rhythm, above all the beautiful

timing of the last two words; and she would, and she does, still find it funny.

The fish surface again in a passage that Gore Vidal calls to our attention, from *Following the Equator*: "'The Whites always mean well when they take human fish out of the ocean and try to make them dry and warm and happy and comfortable in a chicken coop,' which is how, through civilization, they did away with many of the original inhabitants. Lack of empathy is a principal theme in Twain's meditations on race and empire."

Indeed, empathy — and its lack — is a principal theme in virtually all of Twain's work, as contributors frequently note. Nat Hentoff quotes the following thoughts from Huck in *Tom Sawyer Abroad*:

> I see a bird setting on a dead limb of a high tree, singing with its head tilted back and its mouth open, and before I thought I fired, and his song stopped and he fell straight down from the limb, all limp like a rag, and I run and picked him up and he was dead, and his body was warm in my hand, and his head rolled about this way and that, like his neck was broke, and there was a little white skin over his eyes, and one little drop of blood on the side of his head; and laws! I could n't see nothing more for the tears; and I hain't never murdered no creature since that war n't doing me no harm, and I ain't going to.[11]

"The Humane Society," Hentoff writes, "has yet to say anything as powerful — and lasting."

Readers of The Oxford Mark Twain will have the pleasure of revisiting Twain's Mississippi landmarks alongside Willie Morris, whose own lower Mississippi Valley boyhood gives him a special sense of connection to Twain. Morris knows firsthand the mosquitoes described in *Life on the Mississippi* — so colossal that "two of them could whip a dog" and "four of them could hold a man down"; in Morris's own hometown they were so large during the flood season that "local wags said they wore wristwatches." Morris's Yazoo City and Twain's Hannibal shared a "rough-hewn democracy . . . complicated by all the visible textures of caste and class, . . . harmless boyhood fun and mischief right along with . . . rank hypocrisies, churchgoing sanctimonies, racial hatred, entrenched and unrepentant greed."

For the West of Mark Twain's *Roughing It*, readers will have George Plimpton as their guide. "What a group these newspapermen were!" Plimpton writes about Twain and his friends Dan De Quille and Joe Goodman in Virginia City, Nevada. "Their roisterous carryings-on bring to mind the kind of frat-house enthusiasm one associates with college humor magazines like the *Harvard Lampoon*." Malcolm Bradbury examines Twain as "a living example of what made the American so different from the European." And Hal Holbrook, who has interpreted Mark Twain on stage for some forty years, describes how Twain "played" during the civil rights movement, during the Vietnam War, during the Gulf War, and in Prague on the eve of the demise of Communism.

Why do we continue to read Mark Twain? What draws us to him? His wit? His compassion? His humor? His bravura? His humility? His understanding of who and what we are in those parts of our being that we rarely open to view? Our sense that he knows we can do better than we do? Our sense that he knows we can't? E. L. Doctorow tells us that children are attracted to *Tom Sawyer* because in this book "the young reader confirms his own hope that no matter how troubled his relations with his elders may be, beneath all their disapproval is their underlying love for him, constant and steadfast." Readers in general, Arthur Miller writes, value Twain's "insights into America's always uncertain moral life and its shifting but everlasting hypocrisies"; we appreciate the fact that he "is not using his alienation from the public illusions of his hour in order to reject the country implicitly as though he could live without it, but manifestly in order to correct it." Perhaps we keep reading Mark Twain because, in Miller's words, he "wrote much more like a father than a son. He doesn't seem to be sitting in class taunting the teacher but standing at the head of it challenging his students to acknowledge their own humanity, that is, their immemorial attraction to the untrue."

Mark Twain entered the public eye at a time when many of his countrymen considered "American culture" an oxymoron; he died four years before a world conflagration that would lead many to question whether the contradiction in terms was not "European civilization" instead. In between he worked in journalism, printing, steamboating, mining, lecturing, publishing, and editing, in virtually every region of the country. He tried his hand at humorous sketches, social satire, historical novels, children's books, poetry, drama, science fiction, mysteries, romance, philosophy, travelogue, memoir, polemic, and several genres no one had ever seen before or has ever seen since. He invented a self-pasting scrapbook, a history game, a vest strap, and a gizmo for keeping bed sheets tucked in; he invested in machines and processes designed to revolutionize typesetting and engraving, and in a food supplement called "Plasmon." Along the way he cheerfully impersonated himself and prior versions of himself for doting publics on five continents while playing out a charming rags-to-riches story followed by a devastating riches-to-rags story followed by yet another great American comeback. He had a long-running real-life engagement in a sumptuous comedy of manners, and then in a real-life tragedy not of his own design: during the last fourteen years of his life almost everyone he ever loved was taken from him by disease and death.

Mark Twain has indelibly shaped our views of who and what the United States is as a nation and of who and what we might become. He understood the nostalgia for a "simpler" past that increased as that past receded — and he saw through the nostalgia to a past that was just as complex as the present. He recognized better than we did ourselves our potential for greatness and our potential for disaster. His fictions brilliantly illuminated the world in which he lived, changing it — and us — in the process. He knew that our feet often danced to tunes that had somehow remained beyond our hearing; with perfect pitch he played them back to us.

My mother read *Tom Sawyer* to me as a bedtime story when I was eleven. I thought Huck and Tom could be a lot of fun, but I dismissed Becky Thatcher as a bore. When I was twelve I invested a nickel at a local garage sale in a book that contained short pieces by Mark Twain. That was where I met Twain's Eve. Now, *that's* more like it, I decided, pleased to meet a female character I could identify *with* instead of against. Eve had spunk. Even if she got a lot wrong, you had to give her credit for trying. "The Man That Corrupted

Hadleyburg" left me giddy with satisfaction: none of my adolescent reveries of getting even with my enemies were half as neat as the plot of the man who got back at that town. "How I Edited an Agricultural Paper" set me off in uncontrollable giggles.

People sometimes told me that I looked like Huck Finn. "It's the freckles," they'd explain — not explaining anything at all. I didn't read *Huckleberry Finn* until junior year in high school in my English class. It was the fall of 1965. I was living in a small town in Connecticut. I expected a sequel to *Tom Sawyer*. So when the teacher handed out the books and announced our assignment, my jaw dropped: "Write a paper on how Mark Twain used irony to attack racism in *Huckleberry Finn*."

The year before, the bodies of three young men who had gone to Mississippi to help blacks register to vote — James Chaney, Andrew Goodman, and Michael Schwerner — had been found in a shallow grave; a group of white segregationists (the county sheriff among them) had been arrested in connection with the murders. America's inner cities were simmering with pent-up rage that began to explode in the summer of 1965, when riots in Watts left thirty-four people dead. None of this made any sense to me. I was confused, angry, certain that there was something missing from the news stories I read each day: the why. Then I met Pap Finn. And the Phelpses.

Pap Finn, Huck tells us, "had been drunk over in town" and "was just all mud." He erupts into a drunken tirade about "a free nigger . . . from Ohio — a mulatter, most as white as a white man," with "the whitest shirt on you ever see, too, and the shiniest hat; and there ain't a man in town that's got as fine clothes as what he had."

> . . . they said he was a p'fessor in a college, and could talk all kinds of languages, and knowed everything. And that ain't the wust. They said he could *vote*, when he was at home. Well, that let me out. Thinks I, what is the country a-coming to? It was 'lection day, and I was just about to go and vote, myself, if I warn't too drunk to get there; but when they told me there was a State in this country where they'd let that nigger vote, I drawed out. I says I'll never vote agin. Them's the very words I said. . . . And to see the

> cool way of that nigger — why, he wouldn't a give me the road if I hadn't shoved him out o' the way.[12]

Later on in the novel, when the runaway slave Jim gives up his freedom to nurse a wounded Tom Sawyer, a white doctor testifies to the stunning altruism of his actions. The Phelpses and their neighbors, all fine, upstanding, well-meaning, churchgoing folk,

> agreed that Jim had acted very well, and was deserving to have some notice took of it, and reward. So every one of them promised, right out and hearty, that they wouldn't curse him no more.
>
> Then they come out and locked him up. I hoped they was going to say he could have one or two of the chains took off, because they was rotten heavy, or could have meat and greens with his bread and water, but they didn't think of it.[13]

Why did the behavior of these people tell me more about why Watts burned than anything I had read in the daily paper? And why did a drunk Pap Finn railing against a black college professor from Ohio whose vote was as good as his own tell me more about white anxiety over black political power than anything I had seen on the evening news?

Mark Twain knew that there was nothing, absolutely *nothing*, a black man could do — including selflessly sacrificing his freedom, the only thing of value he had — that would make white society see beyond the color of his skin. And Mark Twain knew that depicting racists with chilling accuracy would expose the viciousness of their world view like nothing else could. It was an insight echoed some eighty years after Mark Twain penned Pap Finn's rantings about the black professor, when Malcolm X famously asked, "Do you know what white racists call black Ph.D.'s?" and answered, "'*Nigger!*'"[14]

Mark Twain taught me things I needed to know. He taught me to understand the raw racism that lay behind what I saw on the evening news. He taught me that the most well-meaning people can be hurtful and myopic. He taught me to recognize the supreme irony of a country founded in freedom that continued to deny freedom to so many of its citizens. Every time I hear of another effort to kick Huck Finn out of school somewhere, I recall everything that Mark Twain taught *this* high school junior, and I find myself jumping into the fray.[15] I remember the black high school student who called CNN during the phone-in portion of a 1985 debate between Dr. John Wallace, a black educator spearheading efforts to ban the book, and myself. She accused Dr. Wallace of insulting her and all black high school students by suggesting they weren't smart enough to understand Mark Twain's irony. And I recall the black cameraman on the *CBS Morning News* who came up to me after he finished shooting another debate between Dr. Wallace and myself. He said he had never read the book by Mark Twain that we had been arguing about—but now he really wanted to. One thing that puzzled him, though, was why a white woman was defending it and a black man was attacking it, because as far as he could see from what we'd been saying, the book made whites look pretty bad.

As I came to understand *Huckleberry Finn* and *Pudd'nhead Wilson* as commentaries on the era now known as the nadir of American race relations, those books pointed me toward the world recorded in nineteenth-century black newspapers and periodicals and in fiction by Mark Twain's black contemporaries. My investigation of the role black voices and traditions played in shaping Mark Twain's art helped make me aware of their role in shaping all of American culture.[16] My research underlined for me the importance of changing the stories we tell about who we are to reflect the realities of what we've been.[17]

Ever since our encounter in high school English, Mark Twain has shown me the potential of American literature and American history to illuminate each other. Rarely have I found a contradiction or complexity we grapple with as a nation that Mark Twain had not puzzled over as well. He insisted on taking America seriously. And he insisted on *not* taking America seriously: "I think that there is but a single specialty with us, only one thing that can be called by the wide name 'American,'" he once wrote. "That is the national devotion to ice-water."[18]

Mark Twain threw back at us our dreams and our denial of those dreams, our greed, our goodness, our ambition, and our laziness, all rattling around together in that vast echo chamber of our talk — that sharp, spunky American talk that Mark Twain figured out how to write down without robbing it of its energy and immediacy. Talk shaped by voices that the official arbiters of "culture" deemed of no importance — voices of children, voices of slaves, voices of servants, voices of ordinary people. Mark Twain listened. And he made us listen. To the stories he told us, and to the truths they conveyed. He still has a lot to say that we need to hear.

Mark Twain lives — in our libraries, classrooms, homes, theaters, movie houses, streets, and most of all in our speech. His optimism energizes us, his despair sobers us, and his willingness to keep wrestling with the hilarious and horrendous complexities of it all keeps us coming back for more. As the twenty-first century approaches, may he continue to goad us, chasten us, delight us, berate us, and cause us to erupt in unrestrained laughter in unexpected places.

NOTES

1. Ernest Hemingway, *Green Hills of Africa* (New York: Charles Scribner's Sons, 1935), 22. George Bernard Shaw to Samuel L. Clemens, July 3, 1907, quoted in Albert Bigelow Paine, *Mark Twain: A Biography* (New York: Harper and Brothers, 1912), 3:1398.

2. Allen Carey-Webb, "Racism and *Huckleberry Finn*: Censorship, Dialogue and Change," *English Journal* 82, no. 7 (November 1993):22.

3. See Louis J. Budd, "Impersonators," in J. R. LeMaster and James D. Wilson, eds., *The Mark Twain Encyclopedia* (New York: Garland Publishing Company, 1993), 389–91.

4. See Shelley Fisher Fishkin, "Ripples and Reverberations," part 3 of *Lighting Out for the Territory: Reflections on Mark Twain and American Culture* (New York: Oxford University Press, 1996).

5. There are two exceptions. Twain published chapters from his autobiography in the *North American Review* in 1906 and 1907, but this material was not published in book form in Twain's lifetime; our volume reproduces the material as it appeared in the *North American Review*. The other exception is our final volume, *Mark Twain's Speeches*, which appeared two months after Twain's death in 1910.

An unauthorized handful of copies of *1601* was privately printed by an Alexander Gunn of Cleveland at the instigation of Twain's friend John Hay in 1880. The first American edition authorized by Mark Twain, however, was printed at the United States Military Academy at West Point in 1882; that is the edition reproduced here.

It should further be noted that four volumes — *The Stolen White Elephant and Other Detective Stories*, *Following the Equator and Anti-imperialist Essays*, *The Diaries of Adam and Eve*, and *1601, and Is Shakespeare Dead?* — bind together material originally published separately. In each case the first American edition of the material is the version that has been reproduced, always in its entirety. Because Twain constantly recycled and repackaged previously published works in his collections of short pieces, a certain amount of duplication is unavoidable. We have selected volumes with an eye toward keeping this duplication to a minimum.

Even the twenty-nine-volume Oxford Mark Twain has had to leave much out. No edition of Twain can ever claim to be "complete," for the man was too prolix, and the file drawers of both ephemera and as yet unpublished texts are deep.

6. With the possible exception of *Mark Twain's Speeches*. Some scholars suspect Twain knew about this book and may have helped shape it, although no hard evidence to that effect has yet surfaced. Twain's involvement in the production process varied greatly from book to book. For a fuller sense of authorial intention, scholars will continue to rely on the superb definitive editions of Twain's works produced by the Mark Twain Project at the University of California at Berkeley as they become available. Dense with annotation documenting textual emendation and related issues, these editions add immeasurably to our understanding of Mark Twain and the genesis of his works.

7. Except for a few titles that were not in its collection. The American Antiquarian Society in Worcester, Massachusetts, provided the first edition of *King Leopold's Soliloquy*; the Elmer Holmes Bobst Library of New York University furnished the 1906–7 volumes of the *North American Review* in which *Chapters from My Autobiography* first appeared; the Harry Ransom Humanities Research Center at the University of Texas at Austin made their copy of the West Point edition of *1601* available; and the Mark Twain Project provided the first edition of *Extract from Captain Stormfield's Visit to Heaven.*

8. The specific copy photographed for Oxford's facsimile edition is indicated in a note on the text at the end of each volume.

9. All quotations from contemporary writers in this essay are taken from their introductions to the volumes of The Oxford Mark Twain, and the quotations from Mark Twain's works are taken from the texts reproduced in The Oxford Mark Twain.

10. *The Diaries of Adam and Eve*, The Oxford Mark Twain [hereafter OMT] (New York: Oxford University Press, 1996), p. 33.

11. *Tom Sawyer Abroad*, OMT, p. 74.

12. *Adventures of Huckleberry Finn*, OMT, p. 49–50.

13. Ibid., p. 358.

14. Malcolm X, *The Autobiography of Malcolm X*, with the assistance of Alex Haley (New York: Grove Press, 1965), p. 284.

15. I do not mean to minimize the challenge of teaching this difficult novel, a challenge for which all teachers may not feel themselves prepared. Elsewhere I have developed some concrete strategies for approaching the book in the classroom, including teaching it in the context of the history of American race relations and alongside books by black writers. See Shelley Fisher Fishkin, "Teaching *Huckleberry Finn*," in James S. Leonard, ed., *Making Mark Twain Work in the Classroom* (Durham: Duke University Press, forthcoming). See also Shelley Fisher Fishkin, *Was Huck Black? Mark Twain and African-American Voices* (New York: Oxford University Press, 1993), pp. 106–8, and a curriculum kit in preparation at the Mark Twain House in Hartford, containing teaching suggestions from myself, David Bradley, Jocelyn Chadwick-Joshua, James Miller, and David E. E. Sloane.

16. See Fishkin, *Was Huck Black?* See also Fishkin, "Interrogating 'Whiteness,' Complicating 'Blackness': Remapping American Culture," in Henry Wonham, ed., *Criticism and the Color Line: Desegregating American Literary Studies* (New Brunswick: Rutgers UP, 1996, pp. 251–90 and in shortened form in *American Quarterly* 47, no. 3 (September 1995):428–66.

17. I explore the roots of my interest in Mark Twain and race at greater length in an essay entitled "Changing the Story," in Jeffrey Rubin-Dorsky and Shelley Fisher Fishkin, eds., *People of the Book: Thirty Scholars Reflect on Their Jewish Identity* (Madison: U of Wisconsin Press, 1996), pp. 47–63.

18. "What Paul Bourget Thinks of Us," *How to Tell a Story and Other Essays*, OMT, p. 197.

INTRODUCTION

Roy Blount Jr.

This is Mark Twain's first book and his own least favorite, his worst-selling and his least read. It has been out of print for well over a century. It embarrassed him. He bought and destroyed the plates. It is also in my opinion his finest book, and I'll tell you why:

I haven't read any of the others.

No, I'm pulling your leg. Here's the real reason this is Mark Twain's finest book: he wrote nearly all of it before he met the Hon. Anson Burlingame, U.S. minister to China.

I don't say his best, I say his finest. Why would I say that if I didn't mean it? You may suspect that I am saying it because this book is arguably more like a book that I myself might have written — I wish — than any of Mark Twain's others. If that is so, it speaks well of my modesty. For *The Celebrated Jumping Frog of Calaveras County, and Other Sketches* is slight.

So was Edith Piaf.

And so have been any number of other artists and artworks that would undoubtedly be far more germane to this discussion than Edith Piaf. But Piaf — the Little Sparrow — was slight. And she has the virtue here of having sprung, unaccountably, to mind.

A virtue shared by the sketches in this book. "Remarkable Instances of Presence of Mind," "A Strange Dream," "An Item Which the Editor Himself Could Not Understand" and the other odd flights of observation and fancy sprang unaccountably to Twain's mind when he was approaching and turning thirty, bumming around the Wild West and contributing to various periodicals. And roughly a hundred and thirty years later they still have that unaccountable bounce.

Springing-to-mind is not the same as off-the-top-of-the-head. Springing-to-mind, whether serendipitous, ambrosial or laboriously feigned, or some of all three, is to off-the-top-of-the-head as a vein of gold is to a handful of cash. A wad of cash might get you in on a vein of gold, but it does not in itself have appreciation potential.

Let's not get bogged down in that investment metaphor, which might require me to argue that the sketches in this book have gotten *better* since the 1860s, an argument hard to make for anyone who wasn't around to read them when they first appeared. Let me just quote a passage, from the title piece, that illustrates what I mean by springing-to-mind.

> Smiley said all a frog wanted was education, and he could do most anything — and I believe him. Why, I've seen him set Dan'l Webster down here on this floor — Dan'l Webster was the name of the frog — and sing out, "Flies, Dan'l, flies!" and quicker'n you could wink, he'd spring straight up, and snake a fly off'n the counter there, and flop down on the floor again as solid as a gob of mud, and fall to scratching the side of his head with his hind foot as indifferent as if he hadn't no idea he'd been doin' any more'n any frog might do. You never see a frog so modest and straightfor'ard as he was, for all he was so gifted. (14–15)

An obvious choice, I suppose. But let me say this: the springiest element of that passage, to my mind, is "solid as a gob of mud." I could go on now at some length about frog-as-gob, frog-as-Anyfrog, frog-in-flight, frog-in-imagination, frog-in-its-own-estimation (excuse me, frog-in-*his*-own-estimation), frog-in-the-speaker's-estimation, frog-in-the-writer's and frog-in-the-reader's, *but I am not going to.*

I might even give that gob of mud a bit more thought and start invoking the Creation Myth, but I am not going to do that either.

Nor will I go into any detail about how lovely is the flow of the serial clauses describing the frog's performance — how those clauses expand like ripples (except that the gob brings us up rather short) in a pool:

. . . he'd spring straight up,
and snake a fly off'n the counter there,
and flop down on the floor again as solid as a gob of mud,
and fall to scratching the side of his head with his hind foot as indifferent as if he hadn't no idea he'd been doin' any more'n any frog might do.

Nor will I dwell on how unselfconsciously fluent that whole passage is: as if Mark Twain had no idea he was doing any more than any yarn-spinner might do. He knew he was good when he wrote that passage; he may even have known that he was better at rendering American vernacular on the page than any writer before him. But he didn't know that he was already just about as good at it as he would ever be, or that he was better at it than any writer after him (to this date) would be, or that he was beginning to lay down the fundamental licks of modern American high-informal narrative.

Nor will I, unless backed to the wall, attempt to prove that the central issue in Mark Twain's work to come is foreshadowed in that passage: how much inner life to ascribe to another living being. (Is Jim in *Huckleberry Finn* property or a person or even something finer? Can a pauper be educated into a prince? Ought children to be parboiled or indulged? Do bluejays think?) The flip side of which is how much soul to assume in oneself.

All I am going to say is this: that whereas almost all of nineteenth-century American humor is now one big gob of mud, Mark Twain's humor still jumps.

Now read this passage, which he wrote *after*, long after, meeting Anson Burlingame, U.S. minister to China.

> The Black Hole was not preserved; it is gone, long, long ago. It is strange. Just as it stood, it was itself a monument; a ready-made one. It was finished, it was complete, its materials were strong and lasting, it needed no furbishing up, no repairs; it merely needed to be let alone. It was the first brick, the Foundation Stone, upon which was reared a mighty Empire — the Indian Empire of Great Britain. It was the ghastly episode of the Black Hole that maddened the British and brought Clive, that young military marvel, raging up from Madras; it was the seed from which sprung Plassey; and it was

that extraordinary battle, whose like had not been seen in the earth since Agincourt, that laid deep and strong the foundations of England's colossal Indian sovereignty.

And yet within the time of men who still live, the Black Hole was torn down and thrown away as carelessly as if its bricks were common clay, not ingots of historic gold. There is no accounting for human beings.

That passage is from *Following the Equator*, published thirty years after *Jumping Frog*. I don't say it is bad—an accomplished humorist may be unfunny, either intentionally or unintentionally, but you can rarely catch one being unintentionally funny. And Mark Twain wrote good clear right-on-along prose as dancers walk gracefully. He goes on, after that passage, to give a vivid brief summary of the Black Hole of Calcutta episode: a hundred and forty-six British prisoners jammed into a bedroom-sized cell overnight in terrible heat, by dawn only twenty-three of them still alive, one of them surviving by sucking the sweat from his sleeves.

But the only thing in that passage that *springs* is, assertedly, Plassey. (According to the *Columbia Encyclopedia*, Plassey was the village where Robert Clive's army won a turning-point battle against the Nawab of Bengal.) That passage is disappointing. Not that we can expect even Mark Twain to extract merriment from the Black Hole, but I think we do expect a sharper irony than the one he draws.

And the great thing about the Mark Twain of *Jumping Frog* is that he *would* have extracted merriment from the Black Hole. Of the twenty-seven sketches in this book, twenty-one mention death, and many of them dwell on it, with near-manic anti-morbidity. "Lucretia Smith's Soldier" gleefully mocks sentimental stories about Civil War casualties—pretty audacious, I should think, considering that it was first published (in a California magazine) in December 1864, just a couple of months after Scarlett O'Hara walked out of an infirmary full of shattered men and beheld a hundred-yard stretch of Decatur Street paved with more wounded and dying, and considering furthermore that Samuel Clemens was a Confederate deserter who avoided the carnage by lighting out for the territory.

Not that I fault his judgment in the matter, but "the Wild Humorist of the Pacific Slope," as he was called at the time of *Jumping Frog*, was an unrepentant traitor to *both* the United and the Confederate States of America. That wild humorist would not have waxed vaguely and staidly ironic about either the Black Hole or the empire that sprang from it. That wild humorist might have made fun, in fact, of a world-traveling sage who is so overmatched by the Black Hole of Calcutta that he focuses oddly on its having been obliterated and then enlivens, sort of, a worked-up history lesson with no more amusing comment than "There is no accounting for human beings."

By the time of *Following the Equator*, life and status had taken their toll on Mark Twain's spirits. Whose spirits don't they? He was still vigorous enough to make an arduous journey and to get a still-enjoyable book out of it.

But this book is finer. *Jumping Frog* is not just the seed from which the Mark Twain empire sprang, it is his most devilish and quicksilver book, the one that makes the fewest concessions (if only because he hadn't worked out what concessions to make) to the book-buying market of his day. It is the work of a man who had profited, artistically, from the company of roughnecks, men of whom the Hon. Anson Burlingame, U.S. minister to China, would not have approved.

All but the last few of this book's sketches were written before Sam Clemens met Burlingame, in Honolulu of all places, in 1866. Burlingame "gave me some advice," Mark Twain wrote toward the end of his life, "which I have never forgotten, and which I have lived by for forty years." The advice was: "Avoid inferiors. Seek your comrades among your superiors in intellect and character; always *climb*."

Had Mark Twain not developed pretensions, he would not have become a major writer. But let us be thankful that he did not meet Mr. Burlingame any sooner. The major writer was at his best while within him a minor persisted, writing out of mischief. If Sam Clemens had commenced to *climb* before he got a good grounding in riffraff and foolery, we wouldn't have this book, nor incidentally would we have *Huckleberry Finn*.

I concede that *Huckleberry Finn* is a better book than this one. But this one is finer, in the sense that a child is finer than a man. *Huckleberry Finn* is important enough that there will always be people arguing strenuously and not unpersuasively that like *Hamlet* and *David Copperfield* and *Crime and Punishment*, it doesn't after all quite *work*. Its resolution is problematic; there are holes in it; it isn't searing enough to do slavery justice; Jim does not get his due; if only Mark Twain had been a more highly conscious artist, *Huckleberry Finn* would be consummate and criticism could but bow in silence.

Whereas in *Jumping Frog*, Mark Twain was having a fine young time. And criticism doesn't have much to sink its teeth into. This introduction is already longer than everything else I have found to read about this book put together, excepting Richard Bucci's afterword to this edition, which I would borrow large chunks from if I thought I could do so undetected.

At this point I will address a consideration that may seem out of place here, but isn't. That consideration, when you get right down to it, is my having contracted to write an introduction that is five thousand words long.

We know from manuscripts that Mark Twain counted his words as he went along. So, with the aid of my computer, do I. Right now — that is to say, as of the end of this sentence — this introduction has attained a length of 2,039 words.

My temptation is to fall back upon a literary device that is not included in *A Dictionary of Literary Devices*, a book written by Bernard Dupriez, translated and adapted by Albert W. Halsall, and published by the University of Toronto Press. I paid $24.95 for it, plus tax. Including the index, it is 545 pages long.

The device I allude to is *padding*.

Padding is a term (as opposed to a substance) seldom encountered, I should think, in rigorous literary criticism nowadays. For one thing, it is not French.* For another thing, it is too useful. (The last two sentences, I would say — only by way of grateful acknowledgment — are Twainian. This sentence and the one just preceding it fall short of that standard.)

As a matter of fact, *A Dictionary of Literary Devices* is translated from the French. But I have had to consult the *Collins-Robert Paperback French-*

*Unlike Edith Piaf.

English English-French Dictionary (Robert-Collins Poche Dictionnaire Français-Anglais Anglais-Français) to learn that the French word for *padding* is *délayage*. I prefer *padding*.

Mark Twain tended to pad his books quite a bit. The subscription publishing market demanded thick books. People had more time to read in those days. (Which scarcely justifies a taste for padding — there is no accounting for human beings.)

But *The Celebrated Jumping Frog of Calaveras County, and Other Sketches* was pared down close to the bone. The book's editor, Charles H. Webb, left out a number of perfectly good sketches, with Mark Twain's perhaps nervous acquiescence, because they did not strike Webb as literary enough for an Eastern audience. There's a rousing parody of boxing journalism called "The Great Prize Fight," which purports to describe an epic free-for-all between the incoming and outgoing governors of California, ending in a heartily applauded draw: the two eminent men "never let up until they had minced each other into such insignificant odds and ends that neither was able to distinguish his own remnants from those of his antagonist. It was awful." I would also have put in a fiercely anti-sentimental piece entitled "Uncle Lige," although its best lines are so disgusting that I won't quote them lest I give you the wrong impression. And at least one of the pieces included in this book, "Answers to Correspondents," was in my view overtrimmed to exclude several nice touches. (The full-blown version of "Answers to Correspondents" is preserved for the ages in the Library of America's *Mark Twain: Collected Tales, Sketches, Speeches, and Essays, 1852–1890*. Also to be found there are "The Great Prize Fight," "Uncle Lige" and a number of other early sketches deemed too rough or ephemeral — or both — for the frog book.) This is probably the only book by Mark Twain that would be even better if it were longer.

I am aware that the same has never been said for *any* introduction.

It remains for me, however, to make more of a case that *Jumping Frog* is Mark Twain's finest book. I am loath to start quoting bits from the sketches, going on and on about how funny those bits are, and explaining why they are so funny. I don't think there's much chance of taking the bloom off, but I expect you would rather do the quoting and the going on and on yourself, and let the explaining slide. There is nothing less useful (or perhaps more French), certainly nothing less funny, than explaining why something is funny.

Here, for example, is a thorough explication of what is so funny about the word *French*:

You don't expect the word for such a stylish culture to rhyme with *wrench*. The word *français* isn't funny, it's about what you would expect: elegant, spirited, nearly rhymes with the way the English say *fancy*. Whereas *French* rhymes with *trench*, *wench*, and *stench*. The word *French* is Anglo-Saxon reprisal against the Norman element in English. (That *French* rhymes with *mensch* is Yiddish lagniappe.) If you have ever heard anyone read aloud from, say, *Beowulf* in the original Anglo-Saxon (something like a Swede intoning Old English while chewing peanut brittle), you will be glad that William the Conqueror did come over and bring some Romance music into the language, but that doesn't mean we have to grant the point to the French.

A sounder proof that the word *French* is funny, however, is that Mark Twain found it so. I can't come up with any textual references, but I recall his writing of some deplorable practice that it was immoral, it was un-American, it was almost . . . French. Part of the joke was that in a more genteel period of American culture *French* was almost synonymous with *racy*, or worse; and yet the word rhymed with *bench*. Of interest in this connection is Mark Twain's freehand literal retranslation of a French translation of the jumping frog story back into English. ("He him had accomplished in the art of to gobble the flies . . . and him there exercised continually — so well that a fly at the most far that she appeared was a fly lost.")

But I am padding. Stealthily, like a burglar in the night, but padding just the same.

Let's look specifically at the fineness of what I consider to be this book's finest sketch, one that I have not seen anywhere else: "An Item Which the Editor Himself Could Not Understand." What points may be made with regard to this sketch? Perhaps the following:

(a.) It is about nothing at all except itself. Its text is itself, its commentary is itself, its reverberations are none but its own. It *reads* itself to itself. It "rides on its own melting," as Robert Frost said poetry should.

(b.) It is more postmodern than anything postmodern I have read.

(c.) It is no dry exercise — is not in fact self-indulgent. Its runaway horse is palpable (as hard to get disentangled from as the runaway sentence in which it appears), and its dead Christian woman breathes.

(d.) It is the funniest thing I have ever read by Mark Twain, except for a hundred other things by Mark Twain, and that is a conservative estimate.

(e.) Excuse me while I give it another reading, purely by way of indulging myself.

The jumping frog sketch, too, is story-within-story. Not to mention Smiley-within-Smiley. It also has one unsmiling narrator within another — and the fit is not quite right. In itself the sabotaging of the frog is not much of a story, but it is *enough* of a story to make the outer narrator's impatience with it seem forced. (In 1894 a professor astonished Mark Twain by showing him a story from ancient Greece whose essentials were the same, except for pebbles in the place of shot. This bit of scholarship was offered and believed in good faith, but it turned out that a textbook on Greek composition had used Twain's tale, translated, for comic relief.) Part of the sketch's layered joke, you might say, is that this formal-English narrator is himself boringly overrefined — a *climber* who would rather hear about someone named Reverend Leonidas W. Smiley than about someone named Jim Smiley — and therefore fails to appreciate Simon Wheeler's pokerface humor. But in fact the outer narrator seems pretty sensible, seems to relate directly to the reader, so it's hard to understand why he wouldn't share with the reader a desire to sit down and hear stories about Jim Smiley all day. His nudging introduction to Simon Wheeler's monologue is too explicit: if I had been editor Webb, I would have cut at least the second and third sentences of the three-sentence first paragraph, the word *monotonous* in the third paragraph, and everything in the third paragraph between *sincerity* and *As I said*.

I would also have hastily struck out the word *good-naturedly* in the sketch's last sentence. Or else asked Mark Twain to rework the outer narrator so that he is more clearly not good-natured. One of the great things about Huck Finn as a narrator is that he doesn't realize how good-natured he is, so that Huck himself and the other characters shimmer with a certain independence. Whereas we seem to be expected to take this narrator's word that the spectacle of Simon Wheeler was "exquisitely absurd," and that the frog-matchers in his story were *not* "men of transcendent genius in finesse."

If you think I'm being picky, consider the author. After this book came out, Mark Twain himself complained of "deadly inconsistencies of spelling in the Frog sketch." I have scoured the frog sketch thoroughly, and I don't find a single inconsistency worth mentioning. Instances not worth mentioning: *doin'* on page 15 is I believe the only instance of a *g* being dropped from a present participle, and I guess you could say that *edercate* is inconsistent with *education* and *throw'd* with *ketched*, and it's true that for *wasn't* there are two *wan'ts* and one *warn't*. In fact when he reprinted this sketch in another book Mark Twain changed *he'd been doin'* to *he'd done, edercate* to *educate*, *throw'd* to *throwed*, one of the *wan'ts* to *warn't* and another one (maybe this was a typesetter's fault) to *wasn't*. He also changed *an't* to *ain't*, *risk* to *resk*, *cal'klate* to *cal'late*, *pints* to *points*, and maybe a couple of other things but my eyes are tired.

Perhaps it grated on the author that the title sketch needed a touch of fine tuning. Or of retuning, after Webb put his hand in. *Wan't* and *an't* sound sort of Bostonian. If an editor from the East indulged his own ear by changing a man's *warn't* and *ain't*, then it's no wonder the man wanted, emotionally, to destroy the plates. But by any reasonable standard, the phonetic spelling as it appears in this book is masterful. It's just flavorful enough to capture Simon's voice (*thish-yer* is a gem), but it's not thick enough to bog the reader down — a great advance in American dialect humor, even over James Russell Lowell's supple but less reader-friendly fidelity to Yankee pronunciation.

The only thing really wrong with this sketch is that Mark Twain — exporting a bit of homegrown Western humor to the East — is *climbing* just a bit, and not entirely sure of his purchase. He has not yet broken free of the stuffy practice (see A. B. Longstreet's *Georgia Scenes*) of making it perfectly —

irritatingly — clear that the author is cultivated enough to know that the funny folk he quotes are barbarous.

Therefore the frog sketch is less fine than "An Item Which the Editor Himself Could Not Understand," in which the author's furious disapproval of the strangely written news item is as funny as, and inseparable from, the item itself. However, the celebrated frog is a more rounded character than the horse in "An Item Which . . ." — in fact he is such a character that I hate to see him sandbagged, so to speak. Any number of formal missteps would be outweighed, so to speak, by a frog that shrugs "like a Frenchman."

And let us not overlook a fighting pup named Andrew Jackson that "would have made a name for hisself if he'd lived." What self-possession Andrew Jackson has! An opponent

> might tackle him, and bully-rag him, and bite him, and throw him over his shoulder two or three times, and Andrew Jackson would never let on but what *he* was satisfied, and hadn't expected nothing else — and the bets being doubled and doubled on the other side all the time. . . . (12)

We can only wish we had learned a little more about Smiley's yaller one-eyed no-tailed cow. Did he race her? Fight her? Leave the reader wanting more.

An admonition that far too few critics take to heart. Of "Aurelia's Unfortunate Young Man," therefore, I will say only that I will take it and you can take, say, *The Prince and the Pauper*, and we'll set them side by side with their forepaws just even, and I'll bet you forty dollars my frog will jump further. Also, it occurs to me that this sketch is a prenuptial variation on an Edgar Allan Poe story entitled "The Man Who Was Used Up," which is not funny *at all*, though it clearly means to be. Why are Poe's comic pieces just bizarre and Mark Twain's hilarious? Maybe death was more of a siren to Poe than it was to Sam Clemens. Mark Twain made many remarks about death's attractiveness, but he mainly cocked a snoot at it. (See, for instance, "Origin of Illustrious Men" and the conclusion of "An Inquiry About Insurances.")

"A Complaint About Correspondents" is flawed by fond condescension toward the child. The author's interjections in Annie's letter are false notes.

When it comes to finesse, you can't hold your own with a child by being nice. (See "Advice for Good Little Girls.")

"Answers to Correspondents." The tasty bits cut out include

> my usual style of ciphering out the merits of poetry, which is to read a line or two near the top, a verse near the bottom and then strike an average

and this description of a rendering, in fresco, of the Great Seal of California (meant to depict the goddess Minerva with a bear, eating grapes, at her feet):

> The female figure is a placid, portly, straight-haired squaw in complete armor, sitting on a recumbent hog, and so absorbed in contemplation of the cobble-stones that she does not observe that she has got her sack of turnips by the wrong end, and that dozens of them are rolling out at the other; neither does she observe that the hog has seized the largest turnip and has got it in his mouth; neither does she observe that her great weight is making it mighty uncomfortable for the hog; she does not notice that she is mashing the breath out of him and making his eyes bulge out with a most agonized expression — nor that it is as much as he can do to hold on to his turnip. There is nothing magnanimous in this picture. Any true-hearted American woman, with the kindly charity and the tenderness that are inseparable from her character, would get up for a minute and give the hog a chance to eat his turnip in peace.

Cut the way it is in this book, I have to say "Answers to Correspondents" is a dud — and not nice *enough* regarding a child. Maybe after all this isn't Mark Twain's finest book. That it is arguably so, however, I have already proved, by arguing so. (That's kind of a Twainian sentence, isn't it? I don't think it's possible, at least for an American writer, to parody Mark Twain. It would be like doing an impression of your father or mother: he or she is already there in your voice.) So now I'm just coasting.

"Among the Fenians": Not a lot to this one.

Oh, but on the other hand "The Story of the Bad Little Boy Who Didn't Come to Grief" is quite nice, and not at all outdated at a time when *The Book of Virtues* sells in the millions.

"Curing a Cold." The result of drinking a quart of warm salt water "was surprising. I believe I threw up my immortal soul." This warrants some commentary, I believe. Somewhere way back up there above, I tossed in the assertion that the central issue of Mark Twain's writing was how much inner life to ascribe to another living being, and on the other hand, how much soul to assume in oneself. Around 1880, in "A Cat Tale," he touched on this issue as follows:

> I proposed to draw a cat recumbent and in repose; but just as I had finished the front end of her, she got up and began to gaze passionately at a bird and wriggle her tail in a most expressively wistful way. I had to finish her with that end standing, and the other end lying. It greatly injures the picture. For, you see, it confuses two passions together — the passion of standing up, and the passion of lying down. These are incompatible; and they convey a bad effect to the picture by rendering it unrestful to the eye. In my opinion a cat in a picture ought to be doing one thing or the other — lying down, or standing up; but not both. I ought to have laid this one down again, and put a brick or something on her; but I did not think of it at the time.

A brick on a cat! What a thought! He loved cats! But then, how about a gravestone on a person? The passion of empathizing with others and the passion of making them be still, the passion of preserving one's spark and the passion of giving up the ghost, the passion of the spring and the passion of the gob of mud, were always confused in Mark Twain. A brick on the cat, a yoke on the slave, propriety on the psyche — it springs to mind, and it palpably won't do.

"Literature in the Dry Diggings" contains the presumable misprint *sashshaying* for *sashaying*. The great thing about a facsimile edition is that we can look at a typo — a gob of ink stuck there on the page, when it should be sashaying — and feel the author looking at it with us and seething.

"'After' Jenkins." Makes you want to read the social notes of Jenkins himself, whoever he was. Ah, Mrs. L.R., with her new false teeth and "enchanting and well sustained smile."

"Among the Spirits." In the mock German *lickterhairowfterfrowlein* (an "astonishing proposition"), might we descry a covert reference to oral sex? (Known today in some quarters as French culture.) Just asking.

"Sketch of George Washington." Wherein the Father of our Country's "moral sublimity" is dragged through various aggregates of brick.

"A Page from a Californian Almanac." In the matter of flying gobs of mud, some good sound advice here: "About this time expect more earthquakes; but do not look for them, on account of the bricks."

"Information for the Million." The source, we now see, of the oft-misquoted "serene confidence which a Christian feels in four aces."

"Launch of the Steamer Capital." If there is a heaven, once a week there will be drinking and laughing at scriptural panoramas with Mark Twain. And his friend M. M. M., etc., Nickerson.

"Concerning Chambermaids." It would be interesting to hear the chambermaids' side of this. "The Broom-Cigar Problem: Tidying Up After Mark Twain." Mark Twain would be the writer for it.

"Remarkable Instances of Presence of Mind" and "The Steed 'Oahu.'" Professors don't want you to enjoy school. So they teach *The Mysterious Stranger* instead of these.

"Short and Singular Rations." Factually interesting, but a strange dying fall to end a humorist's book on. (I reject the notion that the title of this sketch should be the title of the book as a whole.) Mark Twain met the survivors of the clipper ship *Hornet* in the company of the Hon. Anson Burlingame. Perhaps this last sketch is a sign that seriousness is growing upon him. Or maybe it is a sign of his fascination with what people and animals will eat. In "The Great Prize Fight," which Webb elected to leave out of this collection, there is a horse that eats children.

In conclusion, I will change my thesis: this would be Mark Twain's finest book *if I'd had a chance to edit it.* Except then it would have been weighed down all these years by my introduction. (So far exactly 5,301 words. I added *exactly* because 5,300 lacked credibility.) Even as it stands, I commend it to you warmly. Thirty-one years after it came out, a man named Carleton, who had passed up a chance to publish it, came up to Mark Twain and said, "I refused a book of yours and for this I stand without competitor as the prize ass of the nineteenth century." He was right.

THE CELEBRATED JUMPING FROG OF CALAVERAS COUNTY,

And other Sketches

The Celebrated
Jumping Frog
By Mark Twain

THE

CELEBRATED JUMPING FROG

OF

CALAVERAS COUNTY,

And other Sketches.

By Mark Twain.

EDITED BY JOHN PAUL.

New-York:

C. H. WEBB, Publisher, 119 & 121 NASSAU ST.

AMERICAN NEWS CO., AGENTS.

1867.

JOHN A. GRAY & GREEN,

PRINTERS,

16 and 18 Jacob St., New-York.

TO

JOHN SMITH,

WHOM I HAVE KNOWN IN DIVERS AND SUNDRY PLACES ABOUT THE WORLD, AND WHOSE MANY AND MANIFOLD VIRTUES DID ALWAYS COMMAND MY ESTEEM, I

Dedicate this Book.

It is said that the man to whom a volume is dedicated, always buys a copy. If this prove true in the present instance, a princely affluence is about to burst upon

THE AUTHOR.

ADVERTISEMENT.

"Mark Twain" is too well known to the public to require a formal introduction at my hands. By his story of the Frog, he scaled the heights of popularity at a single jump, and won for himself the *sobriquet* of The Wild Humorist of the Pacific Slope. He is also known to fame as The Moralist of the Main; and it is not unlikely that as such he will go down to posterity. It is in his secondary character, as humorist, however, rather than in the primal one of moralist, that I aim to present him in the present volume. And here a ready explanation will be found for the somewhat fragmentary character of many of these sketches; for it was necessary to snatch threads of humor wherever they could be found—very often detaching them from serious articles and moral essays with which they were woven and entangled. Originally written for newspaper publication, many of the articles referred to events of the day, the interest of which has now passed away, and contained local allusions, which the general reader would fail to understand; in such cases excision became imperative. Further than this, remark or comment is unnecessary. Mark Twain never resorts to tricks of spelling nor rhetorical buffoonery for the purpose of provoking a laugh; the vein of his humor runs too rich and deep to make surface-gilding necessary. But there are few who can resist the quaint similes, keen satire, and hard good sense which form the staple of his writings. J. P.

CONTENTS.

THE CELEBRATED JUMPING FROG

OF

CALAVERAS COUNTY.

IN compliance with the request of a friend of mine, who wrote me from the East, I called on good-natured, garrulous old Simon Wheeler, and inquired after my friend's friend, *Leonidas W.* Smiley, as requested to do, and I hereunto append the result. I have a lurking suspicion that *Leonidas W.* Smiley is a myth; that my friend never knew such a personage; and that he only conjectured that, if I asked old Wheeler about him, it would remind him of his infamous *Jim* Smiley, and he would go to work and bore me nearly to death with some infernal reminiscence of him as long and tedious as it should be useless to me. If that was the design, it certainly succeeded.

I found Simon Wheeler dozing comfortably by the bar-room stove of the old, dilapidated tavern in the ancient mining camp of Angel's, and I noticed that he was fat and bald-headed, and had an expression of winning gentleness and simplicity upon his tranquil countenance. He roused up and gave me good-day. I told him a friend of mine had commissioned me to make some inquiries about a cherished companion of his boyhood named *Leonidas W.* Smiley—*Rev. Leonidas W.* Smiley—a young minister of the Gospel, who he had heard was at one time a resident of Angel's Camp. I added that, if Mr. Wheeler could tell me any thing about this Rev. Leonidas W. Smiley, I would feel under many obligations to him.

Simon Wheeler backed me into a corner and blockaded me there with his chair, and then sat me down and reeled off the monotonous narrative which follows this paragraph. He never smiled, he never frowned, he never changed his voice from the gentle-flowing key to which he tuned the initial sentence, he never betrayed the slightest suspicion of enthusiasm; but all through the interminable narrative there ran a vein of impressive earnestness and sincerity, which showed me plainly that, so far from his imagining that there was any thing ridiculous or funny about his story, he regarded it as a really important matter, and admired its two heroes as men of transcendent genius in *finesse.* To me, the spectacle of a man drifting serenely along through such a queer yarn without ever smiling, was exquisitely absurd. As I said before, I asked him to tell me what he knew of Rev. Leonidas W. Smiley, and he replied as follows. I let him go on in his own way, and never interrupted him once:

There was a feller here once by the name of *Jim* Smiley, in the winter of '49—or may be it was the spring of '50—I don't recollect exactly, somehow, though what makes me think it was one or the other is because I remember the big flume wasn't finished when he first came to the camp; but any way, he was the curiosest man about always betting on any thing that turned up you ever see, if he could get any body to bet on the other side; and if he couldn't, he'd change sides. Any way that suited the other man would suit him—any way just so's he got a bet, *he* was satisfied. But still he was lucky, uncommon lucky; he most always come out winner. He was always ready and laying for a chance; there couldn't be no solitry thing mentioned but that feller'd offer to bet on it, and take any side you please, as I was just telling you. If there was a horse-race, you'd find him flush, or you'd find him busted at the end of it; if there was a dog-fight, he'd bet on it; if there was a cat-fight, he'd bet on it; if there was a chicken-fight, he'd bet on it; why, if there was two birds setting on a fence, he would bet you which one would fly first; or if there was a camp-meeting, he would be there reg'lar, to bet on Parson Walker, which he judged to be the best exhorter about here, and so he was, too, and a good man. If he even seen a straddle-bug start to go anywheres, he would bet you how long it would take him to get wherever he was going to, and if you took him up, he would foller that straddle-bug to Mexico but what he would find out where he was bound for and how long he was on the road. Lots of the boys here has seen that Smiley, and can tell you about him. Why, it never made no difference to *him*—he would bet on *any* thing—the dangdest feller. Parson Walker's wife laid very sick once, for a good while, and it seemed as if they warn't going to save her; but one morning he come in, and Smiley asked how she was, and he said she was considerable better—thank the Lord for his inf'nit mercy — and coming on so smart that, with the blessing of Prov'dence, she'd get well yet; and Smiley, before he thought, says, "Well, I'll risk two-and-a-half that she don't, any way."

Thish-yer Smiley had a mare — the boys called her the fifteen-minute nag, but that was only in fun, you know, because, of course, she was faster than that — and he used to win money on that horse, for all she was so slow and always had the asthma, or the distemper, or the consumption, or something of that kind. They used to give her two or three hundred yards start, and then pass her under way; but always at the fag-end of the race she'd get excited and desperate-like, and come cavorting and straddling up, and scattering her legs around limber, sometimes in the air, and sometimes out to one side amongst the fences, and kicking up m-o-r-e dust, and raising m-o-r-e racket with her coughing and sneezing and blowing her nose—and always fetch up at the stand just about a neck ahead, as near as you could cipher it down.

And he had a little small bull pup, that to look at him you'd think he wan't worth a cent, but to set around and look ornery, and lay for a chance to steal something. But as soon as money was up on him, he was a different dog; his under-jaw'd begin to stick out like the fo'castle of a steamboat, and his teeth would uncover, and shine savage like the furnaces. And a dog might tackle him, and bully-rag him, and bite him, and throw him over his shoulder two or three times, and Andrew Jackson—which was the name of the pup—Andrew Jackson would never let on but what *he* was satisfied, and hadn't expected nothing else—and the bets being doubled and doubled on the other side all the time, till the money was all up; and then all of a sudden he would grab that other dog jest by the j'int of his hind leg and freeze to it—not chaw, you understand, but only jest grip and hang on till they throwed up the sponge, if it was a year. Smiley always come out winner on that pup, till he harnessed a dog once that didn't have no hind legs, because they'd been sawed off by a circular saw, and when the thing had gone along far enough, and the money was all up, and he come to make a snatch for his pet holt, he saw in a minute how he'd been imposed on, and how the other dog had him in the door, so to speak, and he 'peared surprised, and then he looked sorter discouraged-like, and didn't try no more to win the fight, and so he got shucked out bad. He give Smiley a look, as much as to say his heart was broke, and it was *his* fault, for putting up a dog that hadn't no hind legs for him to take holt of, which was his main dependence in a fight, and then he limped off a piece and laid down and died. It was a good pup, was that Andrew Jackson, and would have made a name for hisself if he'd lived, for the stuff was in him, and he had genius — I know it, because he hadn't had no opportunities to speak of, and it don't stand to reason that a dog could make such a fight as he could under them circumstances, if he hadn't no talent. It always makes me feel sorry when I think of that last fight of his'n, and the way it turned out.

Well, thish-yer Smiley had rat-tarriers, and chicken cocks, and tom-cats, and all them kind of things, till you couldn't rest, and you couldn't fetch nothing for him to bet on but he'd match you. He ketched a frog one day, and took him home, and said he cal'klated to edercate him; and so he never done nothing for three months but set in his back yard and learn that frog to jump. And you bet you he *did* learn him, too. He'd give him a little punch behind, and the next minute you'd see that frog whirling in the air like a doughnut—see him turn one summerset,• or may be a couple, if he got a good start, and come down flat-footed and all right, like a cat. He got him up so in the matter of catching flies, and kept him in practice so constant, that he'd nail a fly every time as far as he could see him. Smiley said all a frog wanted was education, and he could do most any thing—and I believe him. Why, I've seen him set Dan'l Webster down here on this floor—Dan'l Webster was the name of the frog—and sing out, "Flies, Dan'l, flies!" and quicker'n you could wink, he'd spring straight up, and snake a fly off'n the counter there, and flop down on the floor again as solid as a gob of mud, and fall to scratching the side of his head with his hind foot as indifferent as if he hadn't no idea he'd been doin' any more'n any frog might do. You never see a frog so modest and straightfor'ard as he was, for all he was so gifted. And when it come to fair and square jumping on a dead level, he could get over more ground at one straddle than any animal of his breed you ever see. Jumping on a dead level was his strong suit, you understand; and when it come to that, Smiley would ante up money on him as long as he had a red. Smiley was monstrous proud of his frog, and well he might be, for fellers that had traveled and been everywheres, all said he laid over any frog that ever *they* see.

Well, Smiley kept the beast in a little lattice box, and he used to fetch him down town sometimes and lay for a bet. One day a feller —a stranger in the camp, he was—come across him with his box, and says:

"What might it be that you've got in the box?"

And Smiley says, sorter indifferent like, "It

might be a parrot, or it might be a canary, may be, but it an't—it's only just a frog."

And the feller took it, and looked at it careful, and turned it round this way and that, and says, "H'm—so 'tis. Well, what's *he* good for?"

"Well," Smiley says, easy and careless, "He's good enough for *one* thing, I should judge—he can outjump ary frog in Calaveras county."

The feller took the box again, and took another long, particular look, and give it back to Smiley, and says, very deliberate, "Well, I don't see no p'ints about that frog that's any better'n any other frog."

"May be you don't," Smiley says. "May be you understand frogs, and may be you don't understand 'em; may be you've had experience, and may be you an't only a amature, as it were. Anyways, I've got *my* opinion, and I'll risk forty dollars that he can outjump any frog in Calaveras county."

And the feller studied a minute, and then says, kinder sad like, "Well, I'm only a stranger here, and I an't got no frog; but if I had a frog, I'd bet you."

And then Smiley says, "That's all right—that's all right—if you'll hold my box a minute, I'll go and get you a frog." And so the feller took the box, and put up his forty dollars along with Smiley's, and set down to wait.

So he set there a good while thinking and thinking to hisself, and then he got the frog out and prized his mouth open and took a teaspoon and filled him full of quail shot—filled him pretty near up to his chin—and set him on the floor. Smiley he went to the swamp and slopped around in the mud for a long time, and finally he ketched a frog, and fetched him in, and give him to this feller, and says:

"Now, if you're ready, set him alongside of Dan'l, with his fore-paws just even with Dan'l, and I'll give the word." Then he says, "One—two—three—jump!" and him and the feller touched up the frogs from behind, and the new frog hopped off, but Dan'l give a heave, and hysted up his shoulders—so—like a Frenchman, but it wan't no use—he couldn't budge; he was planted as solid as an anvil, and he couldn't no more stir than if he was anchored out. Smiley was a good deal surprised, and he was disgusted too, but he didn't have no idea what the matter was, of course.

The feller took the money and started away; and when he was going out at the door, he sorter jerked his thumb over his shoulders—this way—at Dan'l, and says again, very deliberate, "Well, *I* don't see no p'ints about that frog that's any better'n any other frog."

Smiley he stood scratching his head and looking down at Dan'l a long time, and at last he says, "I do wonder what in the nation that frog throw'd off for—I wonder if there an't something the matter with him—he 'pears to look mighty baggy, somehow." And he ketched Dan'l by the nap of the neck, and lifted him up and says, "Why, blame my cats, if he don't weigh five pound!" and turned him upside down, and he belched out a double handful of shot. And then he see how it was, and he was the maddest man—he set the frog down and took out after that feller, but he never ketched him. And——

[Here Simon Wheeler heard his name called from the front yard, and got up to see what was wanted.] And turning to me as he moved away, he said: "Just set where you are, stranger, and rest easy—I an't going to be gone a second."

But, by your leave, I did not think that a continuation of the history of the enterprising vagabond *Jim* Smiley would be likely tó afford me much information concerning the Rev. *Leonidas W.* Smiley, and so I started away.

At the door I met the sociable Wheeler returning, and he buttonholed me and recommenced:

"Well, thish-yer Smiley had a yaller one-eyed cow that didn't have no tail, only jest a short stump like a bannanner, and——"

"Oh! hang Smiley and his afflicted cow!" I muttered, good-naturedly, and bidding the old gentleman good-day, I departed.

AURELIA'S UNFORTUNATE YOUNG MAN.

THE facts in the following case come to me by letter from a young lady who lives in the beautiful city of San José; she is perfectly unknown to me, and simply signs herself "Aurelia Maria," which may possibly be a fictitious name. But no matter, the poor girl is almost heart-broken by the misfortunes she has undergone, and so confused by the conflicting counsels of misguided friends and insidious enemies, that she does not know what course to pursue in order to extricate herself from the web of difficulties in which she seems almost hopelessly involved. In this dilemma she turns to me for help, and supplicates for my guidance and instruction with a moving eloquence that would touch the heart of a statue. Hear her sad story:

She says that when she was sixteen years old she met and loved, with all the devotion of a passionate nature, a young man from New-Jersey, named Williamson Breckinridge Caruthers, who was some six years her senior. They were engaged, with the free consent of their friends and relatives, and for a time it seemed as if their career was destined to be characterized by an immunity from sorrow beyond the usual lot of humanity. But at last the tide of fortune turned; young Caruthers became infected with small-pox of the most virulent type, and when he recovered from his illness, his face was pitted like a waffle-mould and his comeliness gone forever. Aurelia thought to break off the engagement at first, but pity for her unfortunate lover caused her to postpone the marriage-day for a season, and give him another trial.

The very day before the wedding was to have taken place, Breckinridge, while absorbed in watching the flight of a balloon, walked into a well and fractured one of his legs, and it had to be taken off above the knee. Again Aurelia was moved to break the engagement, but again love triumphed, and she set the day forward and gave him another chance to reform.

And again misfortune overtook the unhappy youth. He lost one arm by the premature discharge of a Fourth-of-July cannon, and within three months he got the other pulled out by a carding-machine. Aurelia's heart was almost crushed by these latter calamities. She could not but be deeply grieved to see her lover passing from her by piecemeal, feeling, as she did, that he could not last forever under this disastrous process of reduction, yet knowing of no way to stop its dreadful career, and in her tearful despair she almost regretted, like brokers who hold on and lose, that she had not taken him at first, before he had suffered such an alarming depreciation. Still, her brave soul bore her up, and she resolved to bear with her friend's unnatural disposition yet a little longer.

Again the wedding-day approached, and again disappointment overshadowed it: Caruthers fell ill with the erysipelas, and lost the use of one of his eyes entirely. The friends and relatives of the bride, considering that she had already put up with more than could reasonably be expected of her, now came forward and insisted that the match should be broken off; but after wavering awhile, Aurelia, with a generous spirit which did her credit, said she had reflected calmly upon the matter, and could not discover that Breckinridge was to blame.

So she extended the time once more, and he broke his other leg.

It was a sad day for the poor girl when she saw the surgeons reverently bearing away the sack whose uses she had learned by previous experience, and her heart told her the bitter truth that some more of her lover was gone. She felt that the field of her affections was growing more and more circumscribed every day, but once more she frowned down her relatives and renewed her betrothal.

Shortly before the time set for the nuptials another disaster occurred. There was but one man scalped by the Owens River Indians last year. That man was Williamson Breckinridge Caruthers, of New-Jersey. He was hurrying home with happiness in his heart, when he lost his hair forever, and in that hour of bitterness he almost cursed the mistaken mercy that had spared his head.

At last Aurelia is in serious perplexity as to what she ought to do. She still loves her Breckinridge, she writes, with true womanly feeling—she still loves what is left of him—but her parents are bitterly opposed to the match, because he has no property and is disabled from working, and she has not sufficient means to support both comfortably. "Now, what should she do?" she asks with painful and anxious solicitude.

It is a delicate question; it is one which involves the lifelong happiness of a woman, and that of nearly two thirds of a man, and I feel that it would be assuming too great a responsibility to do more than make a mere suggestion in the case. How would it do to build to him? If Aurelia can afford the expense, let her furnish her mutilated lover with wooden arms and wooden legs, and a glass eye and a wig, and give him another show; give him ninety days, without grace, and if he does not break his neck in the mean time, marry him and take the chances. It does not seem to me that there is much risk, any way, Aurelia, because if he sticks to his infernal propensity for damaging himself every time he sees a good opportunity, his next experiment is bound to finish him, and then you are all right, you know, married or single. If married, the wooden legs and such other valuables as he may possess, revert to the widow, and you see you sustain no actual loss save the cherished fragment of a noble but most unfortunate husband, who honestly strove to do right, but whose extraordinary instincts were against him. Try it, Maria! I have thought the matter over carefully and well, and it is the only chance I see for you. It would have been a happy conceit on the part of Caruthers if he had started with his neck and broken that first; but since he has seen fit to choose a different policy and string himself out as long as possible, I do not think we ought to upbraid him for it if he has enjoyed it. We must do the best we can under the circumstances, and try not to feel exasperated at him.

A COMPLAINT ABOUT CORRESPONDENTS, DATED IN SAN FRANCISCO.

WHAT do you take us for, on this side of the continent? I am addressing myself personally, and with asperity, to every man, woman, and child east of the Rocky Mountains. How do you suppose our minds are constituted, that you will write us such execrable letters—such poor, bald, uninteresting trash? You complain that by the time a man has been on the Pacific coast six months, he seems to lose all concern about matters and things and people in the distant East, and ceases to answer the letters of his friends and even his relatives. It is your own fault. You need a lecture on the subject—a lecture which ought to read about as follows:

There is only one brief, solitary law for letter-writing, and yet you either do not know that law, or else you are so stupid that you never think of it. It is very easy and simple: Write only about things and people your correspondent takes a living interest in.

Can not you remember that law, hereafter, and abide by it? If you are an old friend of the person you are writing to, you know a number of his acquaintances, and you can rest satisfied that even the most trivial things you can write about them will be read with avidity out here on the edge of sunset.

Yet how *do* you write?—how do the most of you write? Why, you drivel and drivel and drivel along in your wooden-headed way about people one never heard of before, and things which one knows nothing at all about and cares less. There is no sense in that. Let me show up your style with a specimen or so. Here is a paragraph from my Aunt Nancy's last letter—received four years ago, and not answered immediately—not at all, I may say:

"St. Louis, 1862.

"Dear Mark: We spent the evening very pleasantly at home yesterday. The Rev. Dr. Macklin and wife, from Peoria, were here. He is an humble laborer in the vineyard, and takes his coffee strong. He is also subject to neuralgia—neuralgia in the head—and is so unassuming and prayerful. There are few such men We had soup for dinner likewise. Although I am not fond of it. O Mark! why *don't* you try to lead a better life? Read II. Kings, from chap. 2 to chap. 24 inclusive. It would be so gratifying to me if you would experience a change of heart. Poor Mrs. Gabrick is dead. You did not know her. She had fits, poor soul. On the 14th the entire army took up the line of march from——"

I always stopped there, because I knew what was coming—the war news, in minute and dry detail—for I could never drive it into those numskulls that the overland telegraph enabled me to know here in San Francisco every day all that transpired in the United States the day before, and that the pony express brought me exhaustive details of all matters pertaining to the war at least two weeks before their letters could possibly reach me. So I naturally skipped their stale war reports, even at the cost of also skipping the inevitable suggestions to read this, that, and the other batch of chapters in the Scriptures, with which they were interlarded at intervals, like snares wherewith to trap the unwary sinner.

Now what was the Rev. Macklin to me? Of what consequence was it to me that he was "an humble laborer in the vineyard," and "took his coffee strong"?—and was "unassuming," and "neuralgic," and "prayerful"? Such a strange conglomeration of virtues could only excite my admiration—nothing more. It could awake no living interest. That there are few such men, and that we had soup for dinner, is simply gratifying—that is all. "Read twenty-two chapters of II. Kings" is a nice shell to fall in the camp of a man who is not studying for the ministry. The intelligence that "poor Mrs. Gabrick" was dead, aroused no enthusiasm—mostly because of the circumstance that I had never heard of her before, I presume. But I was glad she had fits—although a stranger.

Don't you begin to understand, now? Don't you see that there is not a sentence in that letter of any interest in the world to me? I had the war news in advance of it; I could get a much better sermon at church when I needed it; I didn't care any thing about poor Gabrick, not knowing deceased; nor yet the Rev. Macklin, not knowing him either. I said to myself, "Here's not a word about Mary Anne Smith—I wish there was; nor about Georgiana Brown, or Zeb Leavenworth, or Sam Bowen, or Strother Wiley—or about any body else I care a straw for." And so, as this letter was just of a pattern with all that went before it, it was not answered, and one useless correspondence ceased.

My venerable mother is a tolerably good correspondent—she is above the average, at any rate. She puts on her spectacles and takes her scissors and wades into a pile of newspapers, and slashes out column after column—editorials, hotel arrivals, poetry, telegraph news, advertisements, novelettes, old jokes, recipes for making pies, cures for "biles"—any thing that comes handy; it don't matter to her; she is entirely impartial; she slashes out a column, and runs her eye down it over her spectacles—(she looks over them because she can't see through them, but she prefers them to her more serviceable ones because they have got gold rims to them)—runs her eye down the column, and says, "Well, it's from a St. Louis paper, any way," and jams it into the envelope along with her letter. She writes about every body I ever knew or ever heard of; but unhappily, she forgets that when she tells me that "J. B. is dead," and that "W. L. is going to marry T. D." and that "B. K. and R. M. and L. P. J. have all gone to New-Orleans to live," it is more than likely that years of absence may have so dulled my recollection of once familiar names, that their unexplained initials will be as unintelligible as Hebrew unto me. She never writes a name in full, and so I never know whom she is talking about. Therefore I have to guess—and this was how it came that I mourned the death of Bill Kribben when I should have rejoiced over the dissolution of Ben Kenfuron. I failed to cipher the initials out correctly.

The most useful and interesting letters we get here from home are from children seven or eight years old. This is petrified truth. Happily they have got nothing to talk about but home, and neighbors, and family—things their betters think unworthy of transmission thousands of miles. They write simply and naturally, and without straining for effect. They tell all they know, and then stop. They seldom deal in abstractions or moral homilies. Consequently their epistles are brief; but, treating as they do of familiar scenes and persons, always entertaining. Now, therefore, if you would learn the art of letter-writing, let a little child teach you. I have preserved a letter from a small girl eight years of age—preserved it as a curiosity, because it was the only letter I ever got from the States that had any information in it. It runs thus:

St. Louis, 1865.

"Uncle Mark, if you was here, I could tell you about Moses in the Bulrushers again, I know it better now. Mr. Sowerby has got his leg broke off a horse. He was riding it on Sunday. Margaret, that's the maid, Margaret has took all the spittoons, and slop-buckets, and old jugs out of your room, because she says she don't think you're ever coming back any more, you been gone so long. Sissy McElroy's mother has got another little baby. She has them all the time. It has got little blue eyes, like Mr. Swimley that boards there, and looks just like him. I have got a new doll, but Johnny Anderson pulled one of its legs out. Miss Doosenberry was here to-day; I give her your picture, but she said she didn't want it. My cat has got more kittens—oh! you can't think—twice as many as Lottie Belden's. And there's one, such a sweet little buff one with a short tail, and I named it for you. All of them's got names now—General Grant, and Halleck, and Moses, and Margaret, and Deuteronomy, and Captain Semmes, and Exodus, and Leviticus, and Horace Greeley—all named but one, and I am saving it because the one that I named for You's been sick all the time since, and I reckon it'll die. [It appears to have been mighty rough on the short-tailed kitten, naming it for me—I wonder how the reserved victim will stand it.] Uncle Mark, I do believe Hattie Caldwell likes you, and I know she thinks you are pretty, because I heard her say nothing couldn't hurt your good looks—nothing at all—she said, even if you was to have the small-pox ever so bad, you would be just as good-looking as you was before. And my ma says she's ever so smart. [Very.] So no more this time, because General Grant and Moses is fighting.

Annie."

This child treads on my toes, in every

other sentence, with a perfect looseness, but in the simplicity of her time of life she doesn't know it.

I consider that a model letter—an eminently readable and entertaining letter, and, as I said before, it contains more matter of interest and more real information than any letter I ever received from the East. I had rather hear about the cats at home and their truly remarkable names, than listen to a lot of stuff about people I am not acquainted with, or read "The Evil Effects of the Intoxicating Bowl," illustrated on the back with a picture of a ragged scalliwag pelting away right and left, in the midst of his family circle, with a junk bottle.

ANSWERS TO CORRESPONDENTS.

MORAL STATISTICIAN." — I don't want any of your statistics. I took your whole batch and lit my pipe with it. I hate your kind of people. You are always ciphering out how much a man's health is injured, and how much his intellect is impaired, and how many pitiful dollars and cents he wastes in the course of ninety-two years' indulgence in the fatal practice of smoking; and in the equally fatal practice of drinking coffee; and in playing billiards occasionally; and in taking a glass of wine at dinner, etc., etc., etc. And you are always figuring out how many women have been burned to death because of the dangerous fashion of wearing expansive hoops, etc., etc., etc. You never see more than one side of the question. You are blind to the fact that most old men in America smoke and drink coffee, although, according to your theory, they ought to have died young; and that hearty old Englishmen drink wine and survive it, and portly old Dutchmen both drink and smoke freely, and yet grow older and fatter all the time. And you never try to find out how much solid comfort, relaxation, and enjoyment a man derives from smoking in the course of a lifetime, (which is worth ten times the money he would save by letting it alone,) nor the appalling aggregate of happiness lost in a lifetime by your kind of people from *not* smoking. Of course you can save money by denying yourself all these little vicious enjoyments for fifty years; but then what can you do with it? What use can you put it to? Money can't save your infinitesimal soul. All the use that money can be put to is to purchase comfort and enjoyment in this life; therefore, as you are an enemy to comfort and enjoyment, where is the use in accumulating cash? It won't do for you to say that you can use it to better purpose in furnishing a good table, and in charities, and in supporting tract societies, because you know yourself that you people who have no petty vices are never known to give away a cent, and that you stint yourselves so in the matter of food that you are always feeble and hungry. And you never dare to laugh in the daytime for fear some poor wretch, seeing you in a good humor, will try to borrow a dollar of you; and in church you are always down on your knees, with your eyes buried in the cushion, when the contribution-box comes around; and you never give the revenue officers a true statement of your income. Now you know all these things yourself, don't you? Very well, then, what is the use of your stringing out your miserable lives to a lean and withered old age? What is the use of your saving money that is so utterly worthless to you? In a word, why don't you go off somewhere and die, and not be always trying to seduce people into becoming as "ornery" and unlovable as you are yourselves, by your ceaseless and villainous "moral statistics"? Now, I don't approve of dissipation, and I don't indulge in it, either; but I haven't a particle of confidence in a man who has no redeeming petty vices whatever, and so I don't want to hear from you any more. I think you are the very same man who read me a long lecture, last week, about the degrading vice of smoking cigars, and then came back, in my absence, with your vile, reprehensible fire-proof gloves on, and carried off my beautiful parlor-stove.

"SIMON WHEELER," *Sonora.*—The following simple and touching remarks and accompanying poem have just come to hand from the rich gold-mining region of Sonora:

To Mr. Mark Twain: The within parson, which I have sot to poetry under the name and style of "He Done His Level Best," was one among the whitest men I ever see, and it an't every man that knowed him that can find it in his heart to say he's glad the poor cuss is busted and gone home to the States. He was here in an early day, and he was the handyest man about takin' holt of any thing that come along you most ever see, I judge. He was a cheerful, stirrin' cretur', always doin' something, and no man can say he ever see him do any thing by halvers. Preachin' was his nateral gait, but he warn't a man to lay back and twidle his thums because there didn't happen to be nothin' doin' in his own espeshial line—no, sir, he was a man who would meander forth and stir up something for hisself. His last acts was to go his pile on "kings-*and,*" (calklatin' to fill, but which he didn't fill,) when there was a "flush" out agin him, and naterally, you see, he went under. And so he was cleaned out, as you may say, and he struck the home-trail, cheerful but flat broke. I knowed this talonted man in Arkansaw, and if you would print this humbly tribute to his gorgis abillities, you would greatly obleege his onhappy friend.

HE DONE HIS LEVEL BEST.

Was he a mining on the flat—
  He done it with a zest;
Was he a leading of the choir—
  He done his level best.

If he'd a reg'lar task to do,
  He never took no rest;
Or if 'twas off-and-on—the same—
  He done his level best.

If he was preachin' on his beat,
  He'd tramp from east to west,
And north to south—in cold and heat
  He done his level best.

He'd yank a sinner outen (Hades),*
  And land him with the blest;
Then snatch a prayer 'n waltz in again,
  And do his level best.

He'd cuss and sing and howl and pray,
  And dance and drink and jest,
And lie and steal—all one to him—
  He done his level best.

Whate'er this man was sot to do,
  He done it with a zest;
No matter *what* his contract was,
  HE'D DO HIS LEVEL BEST.

Verily, this man *was* gifted with "gorgis abillities," and it is a happiness to me to embalm the memory of their lustre in these columns. If it were not that the poet crop is unusually large and rank in California this year, I would encourage you to continue writing, Simon; but as it is, perhaps it might be too

* Here I have taken a slight liberty with the original MS. "Hades" does not make such good metre as the other word of one syllable, but it sounds better.

risky in you to enter against so much opposition.

"INQUIRER" wishes to know which is the best brand of smoking tobacco, and how it is manufactured. The most popular—mind, I do not feel at liberty to give an opinion as to the best, and so I simply say the most popular—smoking tobacco is the miraculous conglomerate they call "Killikinick." It is composed of equal parts of tobacco stems, chopped straw, "old soldiers," fine shavings, oak leaves, dog-fennel, corn-shucks, sunflower petals, outside leaves of the cabbage plant, and any refuse of any description whatever that costs nothing and will burn. After the ingredients are thoroughly mixed together, they are run through a chopping-machine and soaked in a spittoon. The mass is then sprinkled with fragrant Scotch snuff, packed into various seductive shapes, labeled "Genuine Killikinick, from the old original manufactory at Richmond," and sold to consumers at a dollar a pound. The choicest brands contain a double portion of "old soldiers," and sell at a dollar and a half. "Genuine Turkish" tobacco contains a treble quantity of "old soldiers," and is worth two or three dollars, according to the amount of service the said "old soldiers" have previously seen. N. B.—This article is preferred by the Sultan of Turkey; his picture and autograph are on the label. Take a handful of "Killikinick," crush it as fine as you can, and examine it closely, and you will find that you can make as good an analysis of it as I have done; you must not expect to discover any particles of genuine tobacco by this rough method, however—to do that, it will be necessary to take your specimen to the mint and subject it to a fire-assay. A good article of cheap tobacco is now made of chopped pine-straw and Spanish moss; it contains one "old soldier" to the ton, and is called "Fine Old German Tobacco."

"PROFESSIONAL BEGGAR."—No; you are not obliged to take greenbacks at par.

* "MELTON MOWBRAY," *Dutch Flat.*—This correspondent sends a lot of doggerel, and says

* This piece of pleasantry, published in a San Francisco paper, was mistaken by the country journals for seriousness, and many and loud were their denunciations of the ignorance of author and editor, in not knowing that the lines in question were "written by Byron."

it has been regarded as very good in Dutch Flat. I give a specimen verse:

> "The Assyrian came down, like a wolf on the fold,
> And his cohorts were gleaming in purple and gold;
> And the sheen of his spears shone like stars on the sea,
> When the blue wave rolls nightly on deep Galilee."

There, that will do. That may be very good Dutch Flat poetry, but it won't do in the metropolis. It is too smooth and blubbery; it reads like buttermilk gurgling from a jug. What the people ought to have is something spirited—something like "Johnny Comes Marching Home." However, keep on practicing, and you may succeed yet. There is genius in you, but too much blubber.

"AMATEUR SERENADER."—Yes, I will give you some advice, and do it with a good deal of pleasure. I live in a neighborhood which is well stocked with young ladies, and consequently I am excruciatingly sensitive upon the subject of serenading. Sometimes I suffer. In the first place, always tune your instruments before you get within three hundred yards of your destination. This will enable you to take your adored unawares, and create a pleasant surprise by launching out at once upon your music. It astonishes the dogs and cats out of their presence of mind, too, so that, if you hurry, you can get through before they have a chance to recover and interrupt you; besides, there is nothing captivating in the sounds produced in tuning a lot of melancholy guitars and fiddles, and neither does a group of able-bodied, sentimental young men so engaged look at all dignified. Secondly, clear your throats and do all the coughing you have got to do before you arrive at the seat of war. I have known a young lady to be ruthlessly startled out of her slumbers by such a sudden and direful blowing of noses and "h'm-h'm-ing" and coughing, that she imagined the house was beleaguered by victims of consumption from the neighboring hospital. Do you suppose the music was able to make her happy after that? Thirdly, don't stand right under the porch and howl, but get out in the middle of the street, or better still, on the other side of it. Distance lends enchantment to the sound. If you have previously transmitted a hint to the lady that she is going to be serenaded, she will understand whom the music is for; besides, if you occupy a neutral position in the middle of the street, may be all the neighbors round will take stock in your serenade, and invite you to take wine with them. Fourthly, don't sing a whole opera through; enough of a thing's enough. Fifthly, don't sing "Lily Dale." The profound satisfaction that most of us derive from the reflection that the girl treated of in that song is dead, is constantly marred by the resurrection of the lugubrious ditty itself by your kind of people. Sixthly, don't let your screaming tenor soar an octave above all the balance of the chorus, and remain there setting every body's teeth on edge for four blocks around; and, above all, don't let him sing a solo; probably there is nothing in the world so suggestive of serene contentment and perfect bliss as the spectacle of a calf chewing a dish-rag; but the nearest approach to it is your reedy tenor, standing apart, in sickly attitude, with head thrown back and eyes uplifted to the moon, piping his distressing solo. Now do not pass lightly over this matter, friend, but ponder it with that seriousness which its importance entitles it to. Seventhly, after you have run all the chickens and dogs and cats in the vicinity distracted, and roused them into a frenzy of crowing, and cackling, and yawling, and caterwauling, put up your dreadful instruments and go home. Eighthly, as soon as you start, gag your tenor—otherwise he will be letting off a screech every now and then, to let the people know he is around. Your amateur tenor is notoriously the most self-conceited of all God's creatures. Tenthly, don't go serenading at all; it is a wicked, unhappy, and seditious practice, and a calamity to all souls that are weary and desire to slumber and would be at rest. Eleventhly and lastly, the father of the young lady in the next block says that if you come prowling around his neighborhood again, with your infamous scraping and tooting and yelling, he will sally forth and deliver you into the hands of the police. As far as I am concerned myself, I would like to have you come, and come often; but as long as the old man is so prejudiced, perhaps you had better serenade mostly in Oakland, or San José, or around there somewhere.

"ST. CLAIR HIGGINS," *Los Angeles.*—"My life is a failure; I have adored, wildly, madly, and she whom I love has turned coldly from me and shed her affections upon another. What would you advise me to do?"

You should shed your affections on another, also—or on several, if there are enough to go round. Also, do every thing you can to make your former flame unhappy. There is an absurd idea disseminated in novels, that the happier a girl is with another man, the happier it makes the old lover she has blighted. Don't allow yourself to believe any such nonsense as that. The more cause that girl finds to regret that she did not marry you, the more comfortable you will feel over it. It isn't poetical, but it is mighty sound doctrine.

"ARITHMETICUS," *Virginia, Nevada.*—"If it would take a cannon ball $3\frac{1}{8}$ seconds to travel four miles, and $3\frac{3}{8}$ seconds to travel the next four, and $3\frac{5}{8}$ seconds to travel the next four, and if its rate of progress continued to diminish in the same ratio, how long would it take it to go fifteen hundred millions of miles?"

I don't know.

"AMBITIOUS LEARNER," *Oakland.*—Yes, you are right—America was not discovered by Alexander Selkirk.

"DISCARDED LOVER."—"I loved, and still love, the beautiful Edwitha Howard, and intended to marry her. Yet, during my temporary absence at Benicia, last week, alas! she married Jones. Is my happiness to be thus blasted for life? Have I no redress?"

Of course you have. All the law, written and unwritten, is on your side. The *intention* and not the *act* constitutes crime—in other words, constitutes the *deed.* If you call your bosom friend a fool, and *intend* it for an insult, it *is* an insult; but if you do it playfully, and meaning no insult, it is *not* an insult. If you discharge a pistol *accidentally*, and kill a man, you can go free, for you have done no murder; but if you try to kill a man, and manifestly *intend* to kill him, but fail utterly to do it, the law still holds that the *intention* constituted the crime, and you are guilty of murder. Ergo, if you had married Edwitha *accidentally*, and without really *intending* to do it, you would not actually be married to her at all, because the *act* of marriage could not be complete without the *intention.* And ergo, in the strict spirit of the law, since you deliberately *intended* to marry Edwitha, and didn't do it, you are married to her all the same—because, as I said before, the *intention* constitutes the crime. It is as clear as day that Edwitha is your wife, and your redress lies in taking a club and mutilating Jones with it as much as you can. Any man has a right to protect his own wife from the advances of other men. But you have another alternative—you were married to Edwitha *first*, because of your deliberate intention, and now you can prosecute her for bigamy, in subsequently marrying Jones. But there is another phase in this complicated case: You *intended* to marry Edwitha, and consequently, according to law, she is your wife—there is no getting around that; but she didn't marry you, and if she *never intended* to marry you, *you are not her husband*, of course. Ergo, in marrying Jones, she was guilty of bigamy, because she was the wife of another man at the time; which is all very well as far as it goes—but then, don't you see, she had no other *husband* when she married Jones, and consequently she was *not* guilty of bigamy. Now, according to this view of the case, Jones married a *spinster*, who was a *widow* at the same time and another man's *wife* at the same time, and yet who had no *husband* and *never had one*, and never had any *intention* of getting married, and therefore, of course, *never had* been married; and by the same reasoning you are a *bachelor*, because you have never been any one's *husband;* and a *married man*, because you have a wife living; and to all intents and purposes a *widower*, because you have been deprived of that wife; and a consummate *ass* for going off to Benicia in the first place, while things were so mixed. And by this time I have got myself so tangled up in the intricacies of this extraordinary case that I shall have to give up any further attempt to advise you—I might get confused and fail to make myself understood. I think I could take up the argument where I left off, and by following it closely awhile, perhaps I could prove to your satisfaction, either that you never existed at all, or that you are dead now, and consequently don't need the faithless Edwitha—I think I could do that, if it would afford you any comfort.

"PERSECUTED UNFORTUNATE."—You say you owe six months' board, and you have no money to pay it with, and your landlord keeps harassing you about it, and you have made all the excuses and explanations possible, and now you are at a loss what to say to him in future. Well, it is a delicate matter to offer advice in a case like this, but your distress impels me to make a suggestion, at least, since I can not venture to do more. When he next importunes you, how would it do to take him impressively by the hand and ask, with simulated emotion, "*Monsieur Jean, votre chien, comme se porte-il?*" Doubtless that is very bad French, but you will find that it will answer just as well as the unadulterated article.

"ARTHUR AUGUSTUS."—No, you are wrong; that is the proper way to throw a brickbat or a tomahawk; but it doesn't answer so well for a bouquet; you will hurt somebody if you keep it up. Turn your nosegay upside down, take it by the stems, and toss it with an upward sweep. Did you ever pitch quoits? that is the idea. The practice of recklessly heaving immense solid bouquets, of the general size and weight of prize cabbages, from the dizzy altitude of the galleries, is dangerous and very reprehensible. Now, night before last, at the Academy of Music, just after Signorina Sconcia had finished that exquisite melody, "The Last Rose of Summer," one of these floral pile-drivers came cleaving down through the atmosphere of applause, and if she hadn't deployed suddenly to the right, it would have driven her into the floor like a shingle-nail. Of course that bouquet was well meant; but how would you have liked to have been the target? A sincere compliment is always grateful to a lady, so long as you don't try to knock her down with it.

"YOUNG MOTHER."—And so you think a baby is a thing of beauty and a joy forever? Well, the idea is pleasing, but not original; every cow thinks the same of its own calf. Perhaps the cow may not think it so elegantly, but still she thinks it, nevertheless. I honor the cow for it. We all honor this touching maternal instinct wherever we find it, be it in the home of luxury or in the humble cow-shed. But really, madam, when I come to examine the matter in all its bearings, I find that the correctness of your assertion does not manifest itself in all cases. A sore-faced baby, with a neglected nose, can not be conscientiously regarded as a thing of beauty; and inasmuch as babyhood spans but three short years, no baby is competent to be a joy "forever." It pains me thus to demolish two thirds of your pretty sentiment in a single sentence; but the position

I hold in this chair requires that I shall not permit you to deceive and mislead the public with your plausible figures of speech. I know a female baby, aged eighteen months, in this city, which can not hold out as a "joy" twenty-four hours on a stretch, let alone "forever." And it possesses some of the most remarkable eccentricities of character and appetite that have ever fallen under my notice. I will set down here a statement of this infant's operations, (conceived, planned, and carried out by itself, and without suggestion or assistance from its mother or any one else,) during a single day; and what I shall say can be substantiated by the sworn testimony of witnesses.

It commenced by eating one dozen large blue-mass pills, box and all; then it fell down a flight of stairs, and arose with a bruised and purple knot on its forehead, after which it proceeded in quest of further refreshment and amusement. It found a glass trinket ornamented with brass-work—mashed up and ate the glass, and then swallowed the brass. Then it drank about twenty drops of laudanum, and more than a dozen table-spoonfuls of strong spirits of camphor. The reason why it took no more laudanum was because there was no more to take. After this it lay down on its back, and shoved five or six inches of a silver-headed whalebone cane down its throat; got it fast there, and it was all its mother could do to pull the cane out again, without pulling out some of the child with it. Then, being hungry for glass again, it broke up several wine glasses, and fell to eating and swallowing the fragments, not minding a cut or two. Then it ate a quantity of butter, pepper, salt, and California matches, actually taking a spoonful of butter, a spoonful of salt, a spoonful of pepper, and three or four lucifer matches at each mouthful. (I will remark here that this thing of beauty likes painted German lucifers, and eats all she can get of them; but she infinitely prefers California matches, which I regard as a compliment to our home manufactures of more than ordinary value, coming, as it does, from one who is too young to flatter.) Then she washed her head with soap and water, and afterward ate what soap was left, and drank as much of the suds as she had room for; after which she sallied forth and took the cow familiarly by the tail, and got kicked heels over head. At odd times during the day, when this joy forever happened to have nothing particular on hand, she put in the time by climbing up on places, and falling down off them, uniformly damaging herself in the operation. As young as she is, she speaks many words tolerably distinctly; and being plain-spoken in other respects, blunt and to the point, she opens conversation with all strangers, male or female, with the same formula, "How do, Jim?" Not being familiar with the ways of children, it is possible that I have been magnifying into matter of surprise things which may not strike any one who is familiar with infancy as being at all astonishing. However, I can not believe that such is the case, and so I repeat that my report of this baby's performances is strictly true; and if any one doubts it, I can produce the child. I will further engage that she will devour any thing that is given her, (reserving to myself only the right to exclude anvils,) and fall down from any place to which she may be elevated, (merely stipulating that her preference for alighting on her head shall be respected, and, therefore, that the elevation chosen shall be high enough to enable her to accomplish this to her satisfaction.) But I find I have wandered from my subject; so, without further argument, I will reiterate my conviction that not *all* babies are things of beauty and joys forever.

"ARITHMETICUS," *Virginia, Nevada.*—"I am an enthusiastic student of mathematics, and it is so vexatious to me to find my progress constantly impeded by these mysterious arithmetical technicalities. Now do tell me what the difference is between geometry and conchology?"

Here *you* come again, with your diabolical arithmetical conundrums, when I am suffering death with a cold in the head. If you could have seen the expression of ineffable scorn that darkened my countenance a moment ago and was instantly split from the center in every direction like a fractured looking-glass by my last sneeze, you never would have written that disgraceful question. Conchology is a science which has nothing to do with mathematics; it relates only to shells. At the same time, however, a man who opens oysters for a hotel, or shells a fortified town, or sucks eggs, is not, strictly speaking, a conchologist—a fine stroke of sarcasm, that, but it will be lost on such an intellectual clam as you. Now compare conchology and geometry together, and you will see what the difference is, and your question will be answered. But don't torture me with any more of your ghastly arithmetical horrors (for I do detest figures any how) until you know I am rid of my cold. I feel the bitterest animosity toward you at this moment—bothering me in this way, when I can do nothing but sneeze and swear and snort pocket-handkerchiefs to atoms. If I had you in range of my nose, now, I would blow your brains out.

"SOCRATES MURPHY."—You speak of having given offense to a gentleman at the opera by *unconsciously* humming an air which the tenor was singing at the time. Now, part of that is a deliberate falsehood. You were not doing it "unconsciously;" no man does such a mean, vulgar, egotistical thing as that unconsciously. You were doing it to "show off;" you wanted the people around you to know you had been to operas before, and to think you were not such an ignorant, self-conceited, supercilious ass as you looked. I can tell you Arizona opera-sharps, any time; you prowl around beer cellars and listen to some howling-dervish of a Dutchman exterminating an Italian air, and then you come into the Academy and prop yourself up against the wall with the stuffy aspect and the imbecile leer of a clothing store dummy, and go to droning along about half an octave below the tenor, and disgusting every body in your neighborhood with your beery strains. [N. B.—If this rough-shod eloquence of mine touches you on a raw spot occasionally, recollect that I am talking for your good, Murphy, and that I am simplifying my language so as to bring it clearly within the margin of your comprehension; it might be gratifying to you to be addressed as if you were an Oxford graduate, but then you wouldn't understand it, you know.] You have got another abominable habit, my sage-brush amateur. When one of those Italian footmen in British uniform comes in and sings, "O tol de rol!—O Signo-o-o-ra!—loango—congo—Venezue-e-e-la! whack fol de rol!" (which means, "O noble madame! here's one of them dukes from the palace, out here, come to borrow a dollar and a half,") you always stand with expanded eyes and mouth, and one pile-driver uplifted, and your sprawling hands held apart in front of your face, like a couple of canvas-covered hams, and when he gets almost through, how you do uncork your pent-up enthusiasm, and applaud with hoof and palm! You have it pretty much to yourself, and then you look sheepish when you find every body staring at you. But how very idiotic you do look when something really fine is sung—you generally keep quiet, then. Never mind, though, Murphy, entire audiences do things at the opera that they have no business to do; for instance, they never let one of those thousand-dollar singers finish—they always break in with their ill-timed applause, just as he or she, as the case may be, is preparing to throw all his or her concentrated sweetness into the final strain, and so all that sweetness is lost. Write me again, Murphy, I shall always be happy to hear from you.

AMONG THE FENIANS.

WISHING to post myself on one of the most current topics of the day, I, Mark, hunted up an old friend, Dennis McCarthy, who is editor of the new Fenian journal in San Francisco, *The Irish People.* I found him sitting on a sumptuous candle-box, in his shirt-sleeves, solacing himself with a whiff at the national *dhudeen* or *caubeen* or whatever they call it—a clay pipe with no stem to speak of. I thought it might flatter him to address him in his native tongue, and so I bowed with considerable grace and said:

"Arrah!"

And he said, "Be jabers!"

"Och hone!" said I.

"Mavourneen dheelish, acushla machree," replied The McCarthy.

"Erin go bragh," I continued with vivacity.

"Asthore!" responded The McCarthy.

"Tare an' ouns!" said I.

"Bhe dha husth; fag a rogarah lums!" said the bold Fenian.

"Ye have me there, be me sowl!" said I, (for I am not "up" in the niceties of the language, you understand; I only know enough of it to enable me to "keep my end up" in an ordinary conversation.)

THE STORY OF THE BAD LITTLE BOY WHO DIDN'T COME TO GRIEF.

ONCE there was a bad little boy, whose name was Jim—though, if you will notice, you will find that bad little boys are nearly always called James in your Sunday-school books. It was very strange, but still it was true, that this one was called Jim.

He didn't have any sick mother, either — a sick mother who was pious and had the consumption, and would be glad to lie down in the grave and be at rest, but for the strong love she bore her boy, and the anxiety she felt that the world would be harsh and cold towards him when she was gone. Most bad boys in the Sunday books are named James, and have sick mothers, who teach them to say, "Now I lay me down," etc., and sing them to sleep with sweet plaintive voices, and then kiss them good-night, and kneel down by the bedside and weep. But it was different with this fellow. He was named Jim, and there wasn't any thing the matter with his mother—no consumption, or any thing of that kind. She was rather stout than otherwise, and she was not pious; moreover, she was not anxious on Jim's account. She said if he were to break his neck, it wouldn't be much loss. She always spanked Jim to sleep, and she never kissed him good-night; on the contrary, she boxed his ears when she was ready to leave him.

Once this little bad boy stole the key of the pantry and slipped in there and helped himself to some jam, and filled up the vessel with tar, so that his mother would never know the difference; but all at once a terrible feeling didn't come over him, and something didn't seem to whisper to him, "Is it right to disobey my mother? Isn't it sinful to do this? Where do bad little boys go who gobble up their good kind mother's jam?" and then he didn't kneel down all alone and promise never to be wicked any more, and rise up with a light, happy heart, and go and tell his mother all about it, and beg her forgiveness, and be blessed by her with tears of pride and thankfulness in her eyes. No; that is the way with all other bad boys in the books; but it happened otherwise with this Jim, strangely enough. He ate that jam, and said it was bully, in his sinful, vulgar way; and he put in the tar, and said that was bully also, and laughed, and observed that "the old woman would get up and snort" when she found it out; and when she did find it out, he denied knowing any thing about it, and she whipped him severely, and he did the crying himself. Every thing about this boy was curious—every thing turned out differently with him from the way it does to the bad Jameses in the books.

Once he climbed up in Farmer Acorn's apple-tree to steal apples, and the limb didn't break, and he didn't fall and break his arm, and get torn by the farmer's great dog, and then languish on a sick bed for weeks, and repent and become good. Oh! no; he stole as many apples as he wanted, and came down all right; and he was all ready for the dog, too, and knocked him endways with a rock when he came to tear him. It was very strange—nothing like it ever happened in those mild little books with marbled backs, and with pictures in them of men with swallow-tailed coats, and bell-crowned hats, and pantaloons that are short in the legs, and women with the waists of their dresses under their arms and no hoops on. Nothing like it in any of the Sunday-school books.

Once he stole the teacher's penknife, and when he was afraid it would be found out, and he would get whipped, he slipped it into George Wilson's cap—poor Widow Wilson's son, the moral boy, the good little boy of the village, who always obeyed his mother, and never told an untruth, and was fond of his lessons and infatuated with Sunday-school. And when the knife dropped from the cap, and poor George hung his head and blushed, as if in conscious guilt, and the grieved teacher charged the theft upon him, and was just in the very act of bringing the switch down upon his trembling shoulders, a white-haired improbable justice of the peace did not suddenly appear in their midst and strike an attitude and say, "spare this noble boy—there stands the cowering culprit! I was passing the school-door at recess, and, unseen myself, I saw the theft committed!" And then Jim didn't get whaled, and the venerable justice didn't read the tearful school a homily, and take George by the hand and say such a boy deserved to be exalted, and then tell him to come and make his home with him, and sweep out the office, and make fires, and run errands, and chop wood, and study law, and help his wife to do household labors, and have all the balance of the time to play, and get forty cents a month, and be happy. No; it would have happened that way in the books, but it didn't happen that way to Jim. No meddling old clam of a justice dropped in to make trouble, and so the model boy George got threshed, and Jim was glad of it; because, you know, Jim hated moral boys. Jim said he was "down on them milksops." Such was the coarse language of this bad, neglected boy.

But the strangest things that ever happened to Jim was the time he went boating on Sunday and didn't get drowned, and that other time that he got caught out in the storm when he was fishing on Sunday, and didn't get struck by lightning. Why, you might look, and look, and look through the Sunday-school books, from now till next Christmas, and you would never come across any thing like this. Oh! no; you would find that all the bad boys who go boating on Sunday invariably get drowned; and all the bad boys who get caught out in storms, when they are fishing on Sunday, infallibly get struck by lightning. Boats with bad boys in them always upset on Sunday, and it always storms when bad boys go fishing on the Sabbath. How this Jim ever escaped is a mystery to me.

This Jim bore a charmed life — that must have been the way of it. Nothing could hurt him. He even gave the elephant in the menagerie a plug of tobacco, and the elephant didn't knock the top of his head off with his trunk. He browsed around the cupboard after essence of peppermint, and didn't make a mistake and drink aqua fortis. He stole his father's gun and went hunting on the Sabbath, and didn't shoot three or four of his fingers off. He struck his little sister on the temple with his fist when he was angry, and she didn't linger in pain through long summer days, and die with sweet words of forgiveness upon her lips that redoubled the anguish of his breaking heart. No; she got over it. He ran off and went to sea at last, and didn't come back and find himself sad and alone in the world, his loved ones sleeping in the quiet churchyard, and the vine-embowered home of his boyhood tumbled down and gone to decay. Ah! no; he came home drunk as a piper, and got into the station-house the first thing.

And he grew up, and married, and raised a large family, and brained them all with an ax one night, and got wealthy by all manner of cheating and rascality, and now he is the infernalest wickedest scoundrel in his native village, and is universally respected, and belongs to the Legislature.

So you see there never was a bad James in the Sunday-school books that had such a streak of luck as this sinful Jim with the charmed lifɜ.

CURING A COLD.

IT is a good thing, perhaps, to write for the amusement of the public, but it is a far higher and nobler thing to write for their instruction, their profit, their actual and tangible benefit. The latter is the sole object of this article. If it prove the means of restoring to health one solitary sufferer among my race, of lighting up once more the fire of hope and joy in his faded eyes, of bringing back to his dead heart again the quick, generous impulses of other days, I shall be amply rewarded for my labor; my soul will be permeated with the sacred delight a Christian feels when he has done a good, unselfish deed.

Having led a pure and blameless life, I am justified in believing that no man who knows me will reject the suggestions I am about to make, out of fear that I am trying to deceive him. Let the public do itself the honor to read my experience in doctoring a cold, as herein set forth, and then follow in my footsteps.

When the White House was burned in Virginia, I lost my home, my happiness, my constitution, and my trunk. The loss of the two first-named articles was a matter of no great consequence, since a home without a mother or a sister, or a distant young female relative in it, to remind you, by putting your soiled linen out of sight and taking your boots down off the mantle-piece, that there are those who think about you and care for you, is easily obtained. And I cared nothing for the loss of my happiness, because, not being a poet, it could not be possible that melancholy would abide with me long.

But to lose a good constitution and a better trunk were serious misfortunes.

On the day of the fire my constitution succumbed to a severe cold caused by undue exertion in getting ready to do something. I suffered to no purpose, too, because the plan I was figuring at for the extinguishing of the fire was so elaborate that I never got it completed until the middle of the following week.

The first time I began to sneeze, a friend told

me to go and bathe my feet in hot water and go to bed. I did so. Shortly afterward, another friend advised me to get up and take a cold shower-bath. I did that also. Within the hour, another friend assured me that it was policy to "feed a cold and starve a fever." I had both. So I thought it best to fill myself up for the cold, and then keep dark and let the fever starve awhile.

In a case of this kind, I seldom do things by halves; I ate pretty heartily; I conferred my custom upon a stranger who had just opened his restaurant that morning; he waited near me in respectful silence until I had finished feeding my cold, when he inquired if the people about Virginia were much afflicted with colds? I told him I thought they were. He then went out and took in his sign. I started down toward the office, and on the way encountered another bosom friend, who told me that a quart of salt water, taken warm, would come as near curing a cold as any thing in the world. I hardly thought I had room for it, but I tried it any how. The result was surprising. I believe I threw up my immortal soul.

Now, as I am giving my experience only for

the benefit of those who are troubled with the distemper I am writing about, I feel that they will see the propriety of my cautioning them against following such portions of it as proved inefficient with me, and acting upon this conviction, I warn them against warm salt water. It may be a good enough remedy, but I think it is too severe. If I had another cold in the head, and there were no course left me but to take either an earthquake or a quart of warm salt water, I would take my chances on the earthquake.

After the storm which had been raging in my stomach had subsided, and no more good Samaritans happening along, I went on borrowing handkerchiefs again and blowing them to atoms, as had been my custom in the early stages of my cold, until I came across a lady who had just arrived from over the plains, and who said she had lived in a part of the country where doctors were scarce, and had from necessity acquired considerable skill in the treatment of simple "family complaints." I knew she must have had much experience, for she appeared to be a hundred and fifty years old.

She mixed a decoction composed of molasses,

aqua fortis, turpentine, and various other drugs, and instructed me to take a wine-glass full of it every fifteen minutes. I never took but one dose; that was enough; it robbed me of all moral principle, and awoke every unworthy impulse of my nature. Under its malign influence my brain conceived miracles of meanness, but my hands were too feeble to execute them; at that time, had it not been that my strength had surrendered to a succession of assaults from infallible remedies for my cold, I am satisfied that I would have tried to rob the graveyard.

Like most other people I often feel mean, and act accordingly; but until I took that medicine I had never reveled in such supernatural depravity and felt proud of it. At the end of two days I was ready to go to doctoring again. I took a few more unfailing remedies, and finally drove my cold from my head to my lungs.

I got to coughing incessantly, and my voice fell below zero; I conversed in a thundering base, two octaves below my natural tone; I could only compass my regular nightly repose by coughing myself down to a state of utter exhaustion, and then the moment I began to talk in my sleep, my discordant voice woke me up again.

My case grew more and more serious every day. Plain gin was recommended ; I took it. Then gin and molasses ; I took that also. Then gin and onions ; I added the onions, and took all three. I detected no particular result, however, except that I had acquired a breath like a buzzard's.

I found I had to travel for my health. I went to Lake Bigler with my reportorial comrade, Wilson. It is gratifying to me to reflect that we traveled in considerable style ; we went in the Pioneer coach, and my friend took all his baggage with him, consisting of two excellent silk handkerchiefs and a daguerreotype of his grandmother. We sailed and hunted and fished and danced all day, and I doctored my cough all night. By managing in this way, I made out to improve every hour in the twenty-four. But my disease continued to grow worse.

A sheet-bath was recommended. I had never refused a remedy yet, and it seemed poor policy to commence then ; therefore I determined to take a sheet-bath, notwithstanding I had no idea what sort of arrangement it was.

It was administered at midnight, and the weather was very frosty. My breast and back were bared, and a sheet (there appeared to be a thousand yards of it) soaked in ice-water was wound around me until I resembled a swab for a Columbiad.

It is a cruel expedient. When the chilly rag touches one's warm flesh, it makes him start with sudden violence and gasp for breath just as men do in the death agony. It froze the marrow in my bones and stopped the beating of my heart. I thought my time had come.

Young Wilson said the circumstance reminded him of an anecdote about a negro who was being baptized, and who slipped from the parson's grasp, and came near being drowned. He floundered around, though, and finally rose up out of the water considerably strangled and furiously angry, and started ashore at once, spouting water like a whale, and remarking, with great asperity, that "One o' dese days some gen'lman's nigger gwyne to git killed wid jes' such dam foolishness as dis!"

Never take a sheet-bath — never. Next to meeting a lady acquaintance, who, for reasons best known to herself, don't see you when she looks at you, and don't know you when she does see you, it is the most uncomfortable thing in the world.

But, as I was saying, when the sheet-bath failed to cure my cough, a lady friend recommended the application of a mustard plaster to my breast. I believe that would have cured me effectually, if it had not been for young Wilson. When I went to bed, I put my mustard plaster—which was a very gorgeous one, eighteen inches square—where I could reach it when I was ready for it. But young Wilson got hungry in the night, and ate it up. I never saw any body have such an appetite; I am confident that lunatic would have eaten me if I had been healthy.

After sojourning a week at Lake Bigler, I went to Steamboat Springs, and beside the steam baths, I took a lot of the vilest medicines that were ever concocted. They would have cured me, but I had to go back to Virginia, where, notwithstanding the variety of new remedies I absorbed every day, I managed to aggravate my disease by carelessness and undue exposure.

I finally concluded to visit San Francisco, and the first day I got there, a lady at the Lick House told me to drink a quart of whisky every twenty-four hours, and a friend at the Occidental recommended precisely the same course. Each advised me to take a quart; that made half a gallon. I did it, and still live.

Now, with the kindest motives in the world, I offer for the consideration of consumptive patients the variegated course of treatment I have lately gone through. Let them try it; if it don't cure them, it can't more than kill them.

AN INQUIRY ABOUT INSURANCES.

COMING down from Sacramento the other night, I found on a center-table in the saloon of the steamboat, a pamphlet advertisement of an Accident Insurance Company. It interested me a good deal, with its General Accidents, and its Hazardous Tables, and Extra-Hazardous furniture of the same description, and I would like to know something more about it. It is a new thing to me. I want to invest if I come to like it. I want to ask merely a few questions of the man who carries on this Accident shop. For I am an orphan.

He publishes this list as accidents he is willing to insure people against.

General accidents include the Traveling Risk, and also all forms of Dislocations, Broken Bones, Ruptures, Tendons, Sprains, Concussions, Crushings, Bruising, Cuts, Stabs, Gunshot Wounds, Poisoned Wounds, Burns and Scalds, Freezing, Bites, Unprovoked Assaults by Burglars, Robbers, or Murderers, the action of Lightning or Sunstroke, the effects of Explosions, Chemicals, Floods, and Earthquakes, Suffocation by Drowning or Choking — where such accidental injury totally disables the person insured from following his usual avocation, or causes death within three months from the time of the happening of the injury.

I want to address this party as follows:

Now, Smith—I suppose likely your name is Smith—you don't know me and I don't know you, but I am willing to be friendly. I am acquainted with a good many of your family—I know John as well as I know any man—and I think we can come to an understanding about your little game without any hard feelings. For instance:

Do you allow the same money on a dog-bite that you do on an earthquake? Do you take special risks for specific accidents?—that is to say, could I, by getting a policy for dog-bites alone, get it cheaper than if I took a chance in your whole lottery? And if so, and supposing I got insured against earthquakes, would you charge any more for San Francisco earthquakes than for those that prevail in places that are better anchored down? And if I had a policy on earthquakes alone, I couldn't collect on a dog-bite, may be, could I?

If a man had such a policy, and an earthquake shook him up and loosened his joints a good deal, but not enough to incapacitate him from engaging in pursuits which did not require him to be tight, wouldn't you pay him some of his pension? I notice you do not mention Biles. How about Biles? Why do you discriminate between Provoked and Unprovoked Assaults by Burglars? If a burglar entered my house at dead of night, and I, in the excitement natural to such an occasion, should forget myself and say something that provoked him, and he should cripple me, wouldn't I get any thing? But if I provoked him by pure accident, I would have you there, I judge; because you would have to pay for the Accident part of it any how, seeing that insuring against accidents is just your strong suit, you know. Now, that item about protecting a man against freezing is good. It will procure you all the custom you want in this country. Because, you understand, the people hereabouts have suffered a good deal from just such climatic drawbacks as that. Why, three years ago, if a man—being a small fish in the matter of money—went over to Washoe, and bought into a good silver mine, they would let that man go on and pay assessments till his purse got down to about thirty-two Fahrenheit, and then the big fish would close in on him and freeze him out. And from that day forth you might consider that man in the light of a bankrupt community; and you would have him down to a spot, too. But if you are ready to insure against that sort of thing, and can stand it, you can give Washoe a fair start. You might send me an agency. Business? Why, Smith, I could get you more business than you could attend to. With such an understanding as that, the boys would all take a chance.

You don't appear to make any particular mention of taking risks on blighted affections. But if you should conclude to do a little business in that line, you might put me down for six or seven chances. I wouldn't mind expense—you might enter it on the extra hazardous. I suppose I would get ahead of you in the long run any how, likely. I have been blighted a good deal in my time.

But now as to those "Effects of Lightning." Suppose the lightning were to strike out at one of your men and miss him, and fetch another party—could that other party come on you for damages? Or could the relatives of the party thus suddenly snaked out of the bright world in the bloom of his youth come on you in case he was crowded for time? as of course he would be, you know, under such circumstances.

You say you have "issued over sixty thousand policies, forty-five of which have proved fatal and been paid for." Now, do you know, Smith, that that looks just a little shaky to me, in a measure? You appear to have it pretty much all your own way, you see. It is all very well for the lucky forty-five that have died "and been paid for," but how about the other fifty-nine thousand nine hundred and fifty-five? You have got their money, haven't you? but somehow the lightning don't seem to strike them and they don't get any chance at you. Won't their families get fatigued waiting for their dividends? Don't your customers drop off rather slow, so to speak?

You will ruin yourself publishing such damaging statements as that, Smith. I tell you as a friend. If you had said that the fifty-nine thousand nine hundred and fifty-five died, and that forty-five lived, you would have issued about four tons of policies the next week. But people are not going to get insured, when you take so much pains to prove that there is such precious little use in it. Good-by Smith!

LITERATURE IN THE DRY DIGGINGS.

ALTHOUGH a resident of San Francisco, I never heard much about the "Art Union Association" of that city until I got hold of some old newspapers during my three months' stay in the Big Tree region of Calaveras county. Up there, you know, they read *every thing*, because in most of those little camps they have no libraries, and no books to speak of, except now and then a patent office report or a prayer-book, or literature of that kind, in a general way, that will hang on and last a good while when people are careful with it, like miners; but as for novels, they pass them around and wear them out in a week or two. Now there was Coon, a nice, bald-headed man at the hotel in Angels' Camp, I asked him to lend me a book, one rainy day; he was silent a moment, and a shade of melancholy flitted across his fine face, and then he said: "Well, I've got a mighty responsible old Webster Unabridged, what there is left of it, but they started her sloshing around and sloshing around and sloshing around the camp before ever I got a chance to read her myself; and next she went to Murphy's, and from there she went to Jackass Gulch, and now she's gone to San Andreas, and I don't expect I'll ever see that book again. But what makes me mad is, that for all they're so handy about keeping her sashshaying around from shanty to shanty and from camp to camp, none of 'em's ever got a good word for her. Now Coddington had her a week, and she was too many for *him*—he couldn't spell the words; he tackled some of them regular busters, tow'rd the middle, you know, and they throwed him; next, Dyer, *he* tried her a jolt, but he couldn't *pronounce* 'em—Dyer can hunt quail or play seven-up as well as any man, understand, but he can't *pronounce* worth a cuss; he used to worry along well enough, though, till he'd flush one of them rattlers with a clatter of syllables as long as a string of sluice-boxes, and then he'd lose his grip and throw up his hand; and so, finally,

Dick Stoker harnessed her, up there at his cabin, and sweated over her and cussed over her and rastled with her for as much as three weeks, night and day, till he got as far as R, and then passed her over to 'Lige Pickerell, and said she was the all-firedest dryest reading that ever *he* struck. Well, well, if she's come back from San Andreas, you can get her, and prospect her, but I don't reckon there's a good deal left of her by this time, though time was when she was as likely a book as any in the State, and as hefty, and had an amount of general information in her that was astonishing, if any of these cattle had known enough to get it out of her." And ex-corporal Coon proceeded cheerlessly to scout with his brush after the straggling hairs on the rear of his head and drum them to the front for inspection and roll-call, as was his usual custom before turning in for his regular afternoon nap.

"AFTER" JENKINS.

A GRAND affair of a ball—the Pioneers'—came off at the Occidental some time ago. The following notes of the costumes worn by the belles of the occasion may not be uninteresting to the general reader, and Jenkins may get an idea therefrom:

Mrs. W. M. was attired in an elegant *pâté de foie gras*, made expressly for her, and was greatly admired.

Miss S. had her hair done up. She was the center of attraction for the gentlemen and the envy of all the ladies.

Miss G. W. was tastefully dressed in a *tout ensemble*, and was greeted with deafening applause wherever she went.

Mrs. C. N. was superbly arrayed in white kid gloves. Her modest and engaging manner accorded well with the unpretending simplicity of her costume, and caused her to be regarded with absorbing interest by every one.

The charming Miss M. M. B. appeared in a thrilling waterfall, whose exceeding grace and volume compelled the homage of pioneers and emigrants alike. How beautiful she was!

The queenly Mrs. L. R. was attractively attired in her new and beautiful false teeth, and the *bon jour* effect they naturally produced was heightened by her enchanting and well sustained smile. The manner of the lady is charmingly pensive and melancholy, and her troops of admirers desired no greater happiness than to get on the scent of her sozodont-sweetened sighs, and track her through her sinuous course among the gay and restless multitude.

Miss R. P., with that repugnance to ostentation in dress, which is so peculiar to her, was attired in a simple white lace collar, fastened with a neat pearl-button solitaire. The fine contrast between the sparkling vivacity of her natural optic and the steadfast attentiveness of her placid glass eye, was the subject of general and enthusiastic remark.

The radiant and sylph-like Mrs. T. wore hoops. She showed to good advantage, and created a sensation wherever she appeared. She was the gayest of the gay.

Miss C. L. B. had her fine nose elegantly enameled, and the easy grace with which she blew it from time to time, marked her as a cultivated and accomplished woman of the world; its exquisitely modulated tone excited the admiration of all who had the happiness to hear it.

Being offended with Miss X. and our acquaintance having ceased permanently, I will take this opportunity of observing to her that it is of no use for her to be slopping off to every ball that takes place, and flourishing around with a brass oyster-knife skewered through her waterfall, and smiling her sickly smile through her decayed teeth, with her dismal pug nose in the air. There is no use in it —she don't fool any body. Every body knows she is old; every body knows she is repaired (you might almost say built) with artificial bones and hair and muscles and things, from the ground up—put together scrap by scrap; and every body knows, also, that all one would have to do would be to pull out her key-pin and she would go to pieces like a Chinese puzzle. There, now, my faded flower, take that paragraph home with you and amuse yourself with it; and if ever you turn your wart of a nose up at me again, I will sit down and write something that will just make you rise up and howl.

LUCRETIA SMITH'S SOLDIER.

I AM an ardent admirer of those nice, sickly war stories which have lately been so popular, and for the last three months I have been at work upon one of that character, which is now completed. It can be relied upon as true in every particular, inasmuch as the facts it contains were compiled from the official records in the War Department at Washington. It is but just, also, that I should confess that I have drawn largely on *Jomini's Art of War*, the *Message of the President and Accompanying Documents*, and sundry maps and military works, so necessary for reference in building a novel like this. To the accommodating Directors of the Overland Telegraph Company I take pleasure in returning my thanks for tendering me the use of their wires at the customary rates. And finally, to all those kind friends who have, by good deeds or encouraging words, assisted me in my labors upon this story of "Lucretia Smith's Soldier," during the past three months, and whose names are too numerous for special mention, I take this method of tendering my sincerest gratitude.

CHAPTER I.

On a balmy May morning in 1861, the little village of Bluemass, in Massachusetts, lay wrapped in the splendor of the newly-risen sun. Reginald de Whittaker, confidential and only clerk in the house of Bushrod & Ferguson, general drygoods and grocery dealers and keepers of the post-office, rose from his bunk under the counter, and shook himself. After yawning and stretching comfortably, he sprinkled the floor and proceeded to sweep it. He had only half finished his task, however, when he sat down on a keg of nails and fell into a reverie. "This is my last day in this shanty," said he. "How it will surprise Lucretia when she hears I am going for a soldier! How proud she will be, the little darling!" He pictured himself in all manner of warlike situations; the hero of a thousand extraordinary adventures; the man of rising fame; the pet of Fortune at last; and beheld himself, finally, returning to his own home, a bronzed and scarred brigadier-general, to cast his honors and his matured and perfect love at the feet of his Lucretia Borgia Smith.

At this point a thrill of joy and pride suffused his system; but he looked down and saw his broom, and blushed. He came toppling down from the clouds he had been soaring among, and was an obscure clerk again, on a salary of two dollars and a half a week.

CHAPTER II.

At eight o'clock that evening, with a heart palpitating with the proud news he had brought for his beloved, Reginald sat in Mr. Smith's parlor awaiting Lucretia's appearance. The moment she entered, he sprang to meet her, his face lighted by the torch of love that was blazing in his head somewhere and shining through, and ejaculated, "Mine own!" as he opened his arms to receive her.

"Sir!" said she, and drew herself up like an offended queen.

Poor Reginald was stricken dumb with astonishment. This chilling demeanor, this angry rebuff, where he had expected the old, tender welcome, banished the gladness from his heart as the cheerful brightness is swept from the landscape when a dark cloud drifts athwart the face of the sun. He stood bewildered a moment, with a sense of goneness on him like one who finds himself suddenly overboard upon a midnight sea, and beholds the ship pass into shrouding gloom, while the dreadful conviction falls upon his soul that he has not been missed. He tried to speak, but his pallid lips refused their office. At last he murmured:

"O Lucretia! what have I done; what is the matter; why this cruel coldness? Don't you love your Reginald any more?"

Her lips curled in bitter scorn, and she replied, in mocking tones:

"Don't I love my Reginald any more? No, I *don't* love my Reginald any more! Go back to your pitiful junk-shop and grab your pitiful yard-stick, and stuff cotton in your ears, so that you can't hear your country shout to you to fall in and shoulder arms. Go!" And then, unheeding the new light that flashed from his eyes, she fled from the room and slammed the door behind her.

Only a moment more! Only a single moment more, he thought, and he could have told her how he had already answered the summons and signed his name to the muster-roll, and all would have been well; his lost bride would have come back to his arms with words of praise and thanksgiving upon her lips. He made a step forward, once, to recall her, but he remembered that he was no longer an effeminate drygoods student, and his warrior soul scorned to sue for quarter. He strode from the place with martial firmness, and never looked behind him.

CHAPTER III.

WHEN Lucretia awoke next morning, the faint music of fife and the roll of a distant drum came floating upon the soft spring breeze, and as she listened the sounds grew more subdued, and finally passed out of hearing. She lay absorbed in thought for many minutes, and then she sighed and said: "Oh! if he were only with that band of fellows, how I could love him!"

In the course of the day a neighbor dropped in, and when the conversation turned upon the soldiers, the visitor said:

'Reginald de Whittaker looked rather down-hearted, and didn't shout when he marched along with the other boys this morning. I expect it's owing to you, Miss Loo, though when I met him coming here yesterday evening to tell you he'd enlisted, he thought you'd like it and be proud of—— Mercy! what in the nation's the matter with the girl?"

Nothing, only a sudden misery had fallen like a blight upon her heart, and a deadly pallor telegraphed it to her countenance. She rose up without a word and walked with a firm step out of the room; but once within the sacred seclusion of her own chamber, her strong will gave way and she burst into a flood of passionate tears. Bitterly she upbraided herself for her foolish haste of the night before, and her harsh treatment of her lover at the very moment that he had come to anticipate the proudest wish of her heart, and to tell her that he had enrolled himself under the battle-flag, and was going forth to fight as *her* soldier. Alas! other maidens would have soldiers in those glorious fields, and be entitled to the sweet pain of feeling a tender solicitude for them, but she would be unrepresented. No soldier in all the vast armies would breathe her name as he breasted the crimson tide of war! She wept again—or, rather, she went on weeping where she left off a moment before. In her bitterness of spirit she almost cursed the precipitancy that had brought all this sorrow upon her young life. "Drat it!" The words were in her bosom, but she locked them there, and closed her lips against their utterance.

For weeks she nursed her grief in silence, while the roses faded from her cheeks. And through it all she clung to the hope that some day the old love would bloom again in Reginald's heart, and he would write to her; but the long summer days dragged wearily along, and still no letter came. The newspapers teemed with stories of battle and carnage, and eagerly she read them, but always with the same result: the tears welled up and blurred the closing lines—the name she sought was looked for in vain, and the dull aching returned to her sinking heart. Letters to the other girls sometimes contained brief mention of him, and presented always the same picture of him—a morose, unsmiling, desperate man, always in the thickest of the fight, begrimed with powder, and moving calm and unscathed through tempests of shot and shell, as if he bore a charmed life.

But at last, in a long list of maimed and killed, poor Lucretia read these terrible words, and fell fainting to the floor: "*R. D. Whittaker, private soldier, desperately wounded!*"

CHAPTER IV.

On a couch in one of the wards of a hospital at Washington lay a wounded soldier; his head was so profusely bandaged that his features were not visible; but there was no mistaking the happy face of the young girl who sat beside him—it was Lucretia Borgia Smith's. She had hunted him out several weeks before, and since that time she had patiently watched by him and nursed him, coming in the morning as soon as the surgeon had finished dressing his wounds, and never leaving him until relieved at nightfall. A ball had shattered his lower jaw, and he could not utter a syllable; through all her weary vigils she had never once been blessed with a grateful word from his dear lips; yet she stood to her post bravely and without a murmur, feeling that when he did get well again she would hear that which would more than reward her for all her devotion.

At the hour we have chosen for the opening of this chapter, Lucretia was in a tumult of happy excitement; for the surgeon had told her that at last her Whittaker had recovered sufficiently to admit of the removal of the bandages from his head, and she was now waiting with feverish impatience for the doctor to come and disclose the loved features to her view. At last he came, and Lucretia, with beaming eyes and fluttering heart, bent over the couch with anxious expectancy. One bandage was removed, then another and another, and lo! the poor wounded face was revealed to the light of day.

"O my own dar——"

What have we here! What is the matter! Alas! it was the face of a stranger!

Poor Lucretia! With one hand covering her upturned eyes, she staggered back with a moan of anguish. Then a spasm of fury distorted her countenance as she brought her fist down with a crash that made the medicine bottles on the table dance again, and exclaimed:

"Oh! confound my cats, if I haven't gone and fooled away three mortal weeks here, snuffling and slobbering over the wrong soldier!"

It was a sad, sad truth. The wretched but innocent and unwitting impostor was R. D., or Richard Dilworthy Whittaker, of Wisconsin, the soldier of dear little Eugenie Le Mulligan, of that State, and utterly unknown to our unhappy Lucretia B. Smith.

Such is life, and the tail of the serpent is over us all. Let us draw the curtain over this melancholy history—for melancholy it must still remain, during a season at least, for the real Reginald de Whittaker has not turned up yet.

THE

KILLING OF JULIUS CÆSAR "LOCALIZED."

BEING THE ONLY TRUE AND RELIABLE ACCOUNT EVER PUBLISHED; TAKEN FROM THE ROMAN "DAILY EVENING FASCES," OF THE DATE OF THAT TREMENDOUS OCCURRENCE.

NOTHING in the world affords a newspaper reporter so much satisfaction as gathering up the details of a bloody and mysterious murder, and writing them up with aggravated circumstantiality. He takes a living delight in this labor of love—for such it is to him—especially if he knows that all the other papers have gone to press, and his will be the only one that will contain the dreadful intelligence. A feeling of regret has often come over me that I was not reporting in Rome when Cæsar was killed—reporting on an evening paper, and the only one in the city, and getting at least twelve hours ahead of the morning paper boys with this most magnificent "item" that ever fell to the lot of the craft. Other events have happened as startling as this, but none that possessed so peculiarly all the characteristics of the favorite "item" of the present day, magnified into grandeur and sublimity by the high rank, fame, and social and political standing of the actors in it. In imagination I have seen myself skirmishing around old Rome, button-holing soldiers, senators, and citizens by turns, and transferring "all the particulars" from them to my note-book; and, better still, arriving at the base of Pompey's statue in time to say persuasively to the dying Cæsar, "Oh! come now, you an't so far gone, you know, but what you could stir yourself up a little and tell a fellow just how this thing happened, if you was a mind to, couldn't you?—now do!" and get the "straight of it" from his own lips, and be envied by the morning paper hounds!

Ah! if I had lived in those days, I would have written up that item gloatingly, and spiced it with a little moralizing here and plenty of blood there; and some dark, shuddering mystery; and praise and pity for some, and misrepresentation and abuse for others, (who did not patronize the paper,) and gory gashes, and notes of warning as to the tendency of the times, and extravagant descriptions of the excitement in the Senate-house and the street, and all that sort of thing.

However, as I was not permitted to report Cæsar's assassination in the regular way, it has at least afforded me rare satisfaction to translate the following able account of it from the original Latin of the *Roman Daily Evening Fasces* of that date—second edition.

"Our usually quiet city of Rome was thrown into a state of wild excitement yesterday by the occurrence of one of those bloody affrays which sicken the heart and fill the soul with fear, while they inspire all thinking men with forebodings for the future of a city where human life is held so cheaply, and the gravest laws are so openly set at defiance. As the result of that affray, it is our painful duty, as public journalists, to record the death of one of our most esteemed citizens — a man whose name is known wherever this paper circulates, and whose fame it has been our pleasure and our privilege to extend, and also to protect from the tongue of slander and falsehood, to the best of our poor ability. We refer to Mr. J. Cæsar, the Emperor-elect.

"The facts of the case, as nearly as our reporter could determine them from the conflicting statements of eye-witnesses, were about as follows: The affair was an election row, of course. Nine tenths of the ghastly butcheries that disgrace the city nowadays grow out of the bickerings and jealousies and animosities engendered by these accursed elections. Rome would be the gainer by it if her very constables were elected to serve a century; for in our experience we have never even been able to choose a dog-pelter without celebrating the event with a dozen knock-downs and a general cramming of the station-house with drunken vagabonds over night. It is said that when the immense majority for Cæsar at the polls in the market was declared the other day, and the crown was offered to that gentleman, even his amazing unselfishness in refusing it three times was not sufficient to save him from the whispered insults of such men as Casca, of the Tenth Ward, and other hirelings of the disappointed candidate, hailing mostly from the Eleventh and Thirteenth and other outside districts, who were overheard speaking ironically and contemptuously of Mr. Cæsar's conduct upon that occasion.

"We are further informed that there are many among us who think they are justified in believing that the assassination of Julius Cæsar was a put-up thing—a cut-and-dried arrangement, hatched by Marcus Brutus and a lot of his hired roughs, and carried out only too faithfully according to the programme. Whether there be good grounds for this suspicion or not, we leave to the people to judge for themselves, only asking that they will read the following account of the sad occurrence carefully and dispassionately before they render that judgment.

"The Senate was already in session, and Cæsar was coming down the street toward the capitol, conversing with some personal friends, and followed, as usual, by a large number of citizens. Just as he was passing in front of Demosthenes & Thucydides's drug-store, he was observing casually to a gentleman, who, our informant thinks, is a fortune-teller, that the Ides of March were come. The reply was, 'Yes, they are come, but not gone yet.' At this moment Artemidorus stepped up and passed the time of day, and asked Cæsar to read a schedule or a tract, or something of the kind, which he had brought for his perusal. Mr. Decius Brutus also said something about an 'humble suit' which *he* wanted read. Artemidorus begged that attention might be paid to his first, because it was of personal consequence to Cæsar. The latter replied that what concerned himself should be read last, or words to that effect. Artemidorus begged and beseeched him to read the paper instantly.* However, Cæsar shook him off, and refused to read any petition in the street. He then entered the capitol, and the crowd followed him.

"About this time the following conversation was overheard, and we consider that, taken in connection with the events which succeeded it,

* Mark that: it is hinted by William Shakespeare, who saw the beginning and the end of the unfortunate affray, that this "schedule" was simply a note discovering to Cæsar that a plot was brewing to take his life.

it bears an appalling significance: Mr. Papilius Lena remarked to George W. Cassius, (commonly known as the 'Nobby Boy of the Third Ward,') a bruiser in the pay of the Opposition, that he hoped his enterprise to-day might thrive; and when Cassius asked, 'What enterprise?' he only closed his left eye temporarily and said with simulated indifference, 'Fare you well,' and sauntered toward Cæsar. Marcus Brutus, who is suspected of being the ringleader of the band that killed Cæsar, asked what it was that Lena had said. Cassius told him, and added in a low tone, '*I fear our purpose is discovered.*'

"Brutus told his wretched accomplice to keep an eye on Lena, and a moment after Cassius urged that lean and hungry vagrant, Casca, whose reputation here is none of the best, to be sudden, for *he feared prevention.* He then turned to Brutus, apparently much excited, and asked what should be done, and swore that either he or Cæsar *should never turn back* — he would kill himself first. At this time Cæsar was talking to some of the back-country members about the approaching fall elections, and paying little attention to what was going on around him. Billy Trebonius got into conversation with the people's friend and Cæsar's —Mark Antony—and under some pretense or other got him away, and Brutus, Decius Casca, Cinna, Metellus Cimber, and others of the gang of infamous desperadoes that infest Rome at present, closed around the doomed Cæsar. Then Metellus Cimber knelt down and begged that his brother might be recalled from banishment, but Cæsar rebuked him for his fawning, sneaking conduct, and refused to grant his petition. Immediately, at Cimber's request, first Brutus and then Cassius begged for the return of the banished Publius ; but Cæsar still refused. He said he could not be moved; that he was as fixed as the North Star, and proceeded to speak in the most complimentary terms of the firmness of that star, and its steady character. Then he said he was like it, and he believed he was the only man in the country that was; therefore, since he was 'constant' that Cimber should be banished, he was also 'constant' that he should stay banished, and he'd be d—d if he didn't keep him so !

"Instantly seizing upon this shallow pretext for a fight, Casca sprang at Cæsar and struck him with a dirk, Cæsar grabbing him by the arm with his right hand, and launching a blow straight from the shoulder with his left, that sent the reptile bleeding to the earth. He then backed up against Pompey's statue, and squared himself to receive his assailants. Cassius and Cimber and Cinna rushed upon him with their daggers drawn, and the former succeeded in inflicting a wound upon his body; but before he could strike again, and before either of the others could strike at all, Cæsar stretched the three miscreants at his feet with as many blows of his powerful fist. By this time the Senate was in an indescribable uproar; the throng of citizens in the lobbies had blockaded the doors in their frantic efforts to escape from the building, the sergeant-at-arms and his assistants were struggling with the assassins, venerable senators had cast aside their encumbering robes, and were leaping over benches and flying down the aisles in wild confusion toward the shelter of the committee-rooms, and a thousand voices were shouting, 'Po-lice! Po-lice!' in discordant tones that rose above the frightful din like shrieking winds above the roaring of a tempest. And amid it all, great Cæsar stood with his back against the statue, like a lion at bay, and fought his assailants weaponless and hand to hand, with the defiant bearing and the unwavering courage which he had shown before on many a bloody field. Billy Trebonius and Caius Legarius struck him with their daggers and fell, as their brother-conspirators before them had fallen. But at last, when Cæsar saw his old friend Brutus step forward, armed with a murderous knife, it is said he seemed utterly overpowered with grief and amazement, and dropping his invincible left arm by his side, he hid his face in the folds of his mantle and received the treacherous blow without an effort to stay the hand that gave it. He only said, '*Et tu, Brute?*' and fell lifeless on the marble pavement.

"We learn that the coat deceased had on when he was killed was the same he wore in his tent on the afternoon of the day he overcame the Nervii, and that when it was removed from the corpse it was found to be cut and gashed in no less than seven different places. There was nothing in the pockets. It will be exhibited at the coroner's inquest, and will be damning proof of the fact of the killing. These latter facts may be relied on, as we get them from Mark Antony, whose position enables him to learn every item of news connected with the one subject of absorbing interest of to-day.

"LATER.—While the coroner was summoning a jury, Mark Antony and other friends of the late Cæsar got hold of the body, and lugged it off to the Forum, and at last accounts Antony and Brutus were making speeches over it and raising such a row among the people that, as we go to press, the chief of police is satisfied there is going to be a riot, and is taking measures accordingly."

AN ITEM WHICH THE EDITOR HIMSELF COULD NOT UNDERSTAND.

OUR esteemed friend, Mr. John William Skae, of Virginia City, walked into the office where we are sub-editor at a late hour last night, with an expression of profound and heartfelt suffering upon his countenance, and, sighing heavily, laid the following item reverently upon the desk, and walked slowly out again. He paused a moment at the door, and seemed struggling to command his feelings sufficiently to enable him to speak, and then, nodding his head toward his manuscript, ejaculated in a broken voice, "Friend of mine —oh! how sad!" and burst into tears. We were so moved at his distress that we did not think to call him back and endeavor to comfort him until he was gone and it was too late. The paper had already gone to press, but knowing that our friend would consider the publication of this item important, and cherishing the hope that to print it would afford a melancholy satisfaction to his sorrowing heart, we stopped the press at once and inserted it in our columns:

DISTRESSING ACCIDENT.—Last evening about 6 o'clock, as Mr. William Schuyler, an old and respectable citizen of South Park, was leaving his residence to go down town, as has been his usual custom for many years, with the exception only of a short interval in the spring of 1850, during which he was confined to his bed by injuries received in attempting to stop a runaway horse by thoughtlessly placing himself directly in its wake and throwing up his hands and shouting, which, if he had done so even a single moment sooner, must inevitably have frightened the animal still more instead of checking its speed, although disastrous enough to himself as it was, and rendered more melancholy and distressing by reason of the presence of his wife's mother, who was there and saw the sad occurrence, notwithstanding it is at least likely, though not necessarily so, that she should be reconnoitering in another direction when incidents occur, not being vivacious and on the lookout, as a general thing, but even the reverse, as her own mother is said to have stated, who is no more, but died in the full hope of a glorious resurrection, upwards of three years ago, aged 86, being a Christian woman and without guile, as it were, or property, in consequence of the fire of 1849, which destroyed every blasted thing she had in the world. But such is life. Let us all take warning by this solemn occurrence, and let us endeavor so to conduct ourselves that when we come to die we can do it. Let us place our hands upon our hearts, and say with earnestness and sincerity that from this day forth we will beware of the intoxicating bowl.—*First Edition of the Californian.*

The boss-editor has been in here raising the very mischief, and tearing his hair and kicking the furniture about, and abusing me like a pickpocket. He says that every time he leaves me in charge of the paper for half an hour, I get imposed upon by the first infant or the first idiot that comes along. And he says that distressing item of Johnny Skae's is nothing but a lot of distressing bosh, and has got no point to it and no sense in it and no information in it, and that there was no earthly necessity for stopping the press to publish it. He says every man he meets has insinuated that somebody about THE CALIFORNIAN office has gone crazy.

Now all this comes of being good-hearted. If I had been as unaccommodating and unsympathetic as some people, I would have told Johnny Skae that I wouldn't receive his communication at such a late hour, and to go to blazes with it; but no, his snuffling distress touched my heart, and I jumped at the chance of doing something to modify his misery. I never read his item to see whether there was any thing wrong about it, but hastily wrote the few lines which preceded it, and sent it to the printers. And what has my kindness done for me? It has done nothing but bring down upon me a storm of abuse and ornamental blasphemy.

Now, I will just read that item myself, and see if there is any foundation for all this fuss. And if there is, the author of it shall hear from me.

. . . . . .

I have read it, and I am bound to admit that it seems a little mixed at a first glance. However, I will peruse it once more.

. . . . . .

I have read it again, and it does really seem a good deal more mixed than ever.

. . . . . .

I have read it over five times, but if I can get at the meaning of it, I wish I may get my just deserts. It won't bear analysis. There are things about it which I can not understand at all. It don't say whatever became of William Schuyler. It just says enough about him to get one interested in his career, and then drops him. Who is William Schuyler, any how, and what part of South Park did he live in, and if he started down-town at six o'clock, did he ever get there, and if he did, did any thing happen to him? Is *he* the individual that met with the "distressing accident"? Considering the elaborate circumstantiality of detail observable in the item, it seems to me that it ought to contain more information than it does. On the contrary, it is obscure—and not only obscure, but utterly incomprehensible. Was the breaking of Mr. Schuyler's leg, fifteen years ago, the "distressing accident" that plunged Mr. Skae into unspeakable grief, and caused him to come up here at dead of night and stop our press to acquaint the world with the unfortunate circumstance? Or did the "distressing accident" consist in the destruction of Schuyler's mother-in-law's property in early times? Or did it consist in the death of that person herself three years ago? (albeit it does not appear that she died by accident.) In a word, what *did* that "distressing accident" consist in? What did that driveling ass of a Schuyler stand *in the wake* of a runaway horse for, with his shouting and gesticulating, if he wanted to stop him? And how the mischief could he get run over by a horse that had already passed beyond him? And what are we to "take warning" by? and how is this extraordinary chapter of incomprehensibilities going to be a "lesson" to us? And above all, what has the "intoxicating bowl" got to do with it, any how? It is not stated that Schuyler drank, or that his wife drank, or that his mother-in-law drank, or that the horse drank—wherefore, then, the reference to the intoxicating bowl? It does seem to me that, if Mr. Skae had let the intoxicating bowl alone himself, he never would have got into so much trouble about this infernal imaginary distressing accident. I have read his absurd item over and over again, with all its insinuating plausibility, until my head swims; but I can make neither head nor tail of it. There certainly seems to have been an accident of some kind or other, but it is impossible to determine what the nature of it was, or who was the sufferer by it. I do not like to do it, but I feel compelled to request that the next time any thing happens to one of Mr. Skae's friends, he will append such explanatory notes to his account of it as will enable me to find out what sort of an accident it was and whom it happened to. I had rather all his friends should die than that I should be driven to the verge of lunacy again in trying to cipher out the meaning of another such production as the above.

AMONG THE SPIRITS.

THERE was a *séance* in town a few nights since. As I was making for it, in company with the reporter of an evening paper, he said he had seen a gambler named Gus Graham shot down in a town in Illinois years ago by a mob, and as he was probably the only person in San Francisco who knew of the circumstance, he thought he would "give the spirits Graham to chaw on awhile." [N. B.—This young creature is a Democrat, and speaks with the native strength and inelegance of his tribe.] In the course of the show he wrote his old pal's name on a slip of paper, and folded it up tightly and put it in a hat which was passed around, and which already had about five hundred similar documents in it. The pile was dumped on the table, and the medium began to take them up one by one and lay them aside, asking, "Is this spirit present? or this? or this?" About one in fifty would rap, and the person who sent up the name would rise in his place and question the defunct. At last a spirit seized the medium's hand and wrote "Gus Graham" backward. Then the medium went skirmishing through the papers for the corresponding name. And that old sport knew his card by the back! When the medium came to it, after picking up fifty others, he rapped! A committeeman unfolded the paper, and it was the right one. I sent for it and got it. It was all right. However, I suppose all Democrats are on sociable terms with the devil. The young man got up and asked:

"Did you die in '51? '52? '53? '54?——"

Ghost—"Rap, rap, rap."

"Did you die of cholera? diarrhea? dysentery? dog-bite? small-pox? violent death?——"

"Rap, rap, rap."

"Were you hanged? drowned? stabbed? shot?——"

"Rap, rap, rap."

"Did you die in Mississippi? Kentucky? New-York? Sandwich Islands? Texas? Illinois?——"

"Rap, rap, rap."

"In Adams county? Madison? Randolph?——"

"Rap, rap, rap."

It was no use trying to catch the departed gambler. He knew his hand, and played it like a major.

About this time a couple of Germans stepped forward, an elderly man and a spry young fellow, cocked and primed for a sensation. They wrote some names. Then young Ollendorff said something which sounded like—

"Ist ein geist hieraus?" [Bursts of laughter from the audience.]

Three raps—signifying that there *was* a geist hieraus.

"Vollen sie schriehen?" [More laughter.]

Three raps.

"Finzig stollen, linsowfterowlickterhairowfterfrowleineruhackfolderol?"

Incredible as it may seem, the spirit cheerfully answered Yes to that astonishing proposition.

The audience grew more and more boisterously mirthful with every fresh question, and they were informed that the performance could not go on in the midst of so much levity. They became quiet.

The German ghost didn't appear to know any thing at all—couldn't answer the simplest questions. Young Ollendorff finally stated some numbers, and tried to get at the time of the spirit's death; it appeared to be considerably mixed as to whether it died in 1811 or 1812, which was reasonable enough, as it had been so long ago. At last it wrote "12."

Tableau! Young Ollendorff sprang to his feet in a state of consuming excitement. He exclaimed:

"Laties und shentlemen! I write de name fon a man vot lifs! Speerit-rabbing dells me he ties in yahr eighteen hoondred und dwelf, but he yoos as live und helty as——"

The Medium—"Sit down, sir!"

Ollendorff—"But I vant to——"

Medium—"You are not here to make speeches, sir—sit down!" [Mr. O. had squared himself for an oration.]

Mr. O. "But de speerit cheat!—dere is no such speerit——" [All this time applause and laughter by turns from the audience.]

Medium—"Take your seat, sir, and I will explain this matter."

And she explained. And in that explanation she let off a blast which was so terrific that I half expected to see young Ollendorff shot up through the roof. She said he had come up there with fraud and deceit and cheating in his heart, and a kindred spirit had come from the land of shadows to commune with him! She was terribly bitter. She said in substance, though not in words, that perdition was full of just such fellows as Ollendorff, and they were ready on the slightest pretext to rush in and assume any body's name, and rap and write and lie and swindle with a perfect looseness whenever they could rope in a living affinity like poor Ollendorff to communicate with! [Great applause and laughter.]

Ollendorff stood his ground with good pluck, and was going to open his batteries again, when a storm of cries arose all over the house, "Get down! Go on! Clear out! Speak on—we'll hear you! Climb down from that platform! Stay where you are! Vamose! Stick to your post—say your say!"

The medium rose up and said if Ollendorff

remained, she would not. She recognized no one's right to come there and insult her by practicing a deception upon her, and attempting to bring ridicule upon so solemn a thing as her religious belief. The audience then became quiet, and the subjugated Ollendorff retired from the platform.

The other German raised a spirit, questioned it at some length in his own language, and said the answers were correct. The medium claimed to be entirely unacquainted with the German language.

Just then a gentleman called me to the edge of the platform and asked me if I were a Spiritualist. I said I was not. He asked me if I were prejudiced. I said not more than any other unbeliever; but I could not believe in a thing which I could not understand, and I had not seen any thing yet that I could by any possibility cipher out. He said, then, that he didn't think I was the cause of the diffidence shown by the spirits, but he knew there was an antagonistic influence around that table somewhere; he had noticed it from the first; there was a painful negative current passing to his sensitive organization from that direction constantly. I told him I guessed it was that other fellow; and I said, Blame a man who was all the time shedding these infernal negative currents! This appeared to satisfy the mind of the inquiring fanatic, and he sat down.

I had a very dear friend, who, I had heard, had gone to the spirit-land, or perdition, or some of those places, and I desired to know something concerning him. There was something so awful, though, about talking with living, sinful lips to the ghostly dead, that I could hardly bring myself to rise and speak. But at last I got tremblingly up and said with a low and trembling voice:

"Is the spirit of John Smith present?"

(You never can depend on these Smiths; you call for one, and the whole tribe will come clattering out of hell to answer you.)

"Whack! whack! whack! whack!"

Bless me! I believe all the dead and damned John Smiths between San Francisco and perdition boarded that poor little table at once! I was considerably set back—stunned, I may say. The audience urged me to go on, however, and I said:

"What did you die of?"

The Smiths answered to every disease and casualty that men can die of.

"Where did you die?"

They answered Yes to every locality I could name while my geography held out.

"Are you happy where you are?"

There was a vigorous and unanimous "No!" from the late Smiths.

"Is it warm there?"

An educated Smith seized the medium's hand and wrote:

"It's no name for it."

"Did you leave any Smiths in that place when you came away!"

"Dead loads of them!"

I fancied I heard the shadowy Smiths chuckle at this feeble joke—the rare joke that there could be live loads of Smiths where all are dead.

"How many Smiths are present?"

"Eighteen millions—the procession now reaches from here to the other side of China."

"Then there are many Smiths in the kingdom of the lost?"

"The Prince Apollyon calls all new comers Smith on general principles; and continues to do so until he is corrected, if he chances to be mistaken."

"What do lost spirits call their dread abode?"

"They call it the Smithsonian Institute."

I got hold of the right Smith at last—the particular Smith I was after—my dear, lost, lamented friend—and learned that he died a violent death. I feared as much. He said his wife talked him to death. Poor wretch!

By and by up started another Smith. A gentleman in the audience said that this was his Smith. So he questioned him, and this Smith said he too died by violence. He had been a good deal tangled in his religious belief, and was a sort of a cross between a Universalist and a Unitarian; has got straightened out and changed his opinions since he left here; said he was perfectly happy. We proceeded to question this talkative and frolicsome old parson. Among spirits I judge he is the gayest of the gay. He said he had no tangible body; a bullet could pass through him and never make a hole; rain could pass through him as through vapor, and not discommode him in the least, (so I suppose he don't know enough to come in when it rains—or don't care enough;) says heaven and hell are simply mental conditions; spirits in the former have happy and contented minds, and those in the latter are torn by remorse of conscience; says as far as he is concerned, he is all right—he is happy; would not say whether he was a very good or a very bad man on earth, (the shrewd old water-proof nonentity! I asked the question so that I might average my own chances for his luck in the other world, but he saw my drift;) says he has an occupation there—puts in his time teaching and being taught; says there are spheres—grades of perfection—he is making very good progress—has been promoted a sphere or so since his matriculation; (I said mentally, "Go slow, old man, go slow, you have got all eternity before you," and he replied not;) he don't know how many spheres there are, (but I suppose there must be millions, because if a man goes galloping through them at the rate this old Universalist is doing, he will get through an infinitude of them by the time he has been there as long as old Sesostris and those ancient mummies; and there is no estimating how high he will get in even the infancy of eternity—I am afraid the old man is scouring along rather too fast for the style of his surroundings, and the length of time he has got on his hands;) says spirits can not feel heat or cold, (which militates somewhat against all my notions of orthodox damnation—fire and brimstone;) says spirits commune with each other by thought—they have no language; says the distinctions of sex are preserved there—and so forth and so on.

The old parson wrote and talked for an hour, and showed by his quick, shrewd, intelligent replies, that he had not been sitting up nights in the other world for nothing; he had been prying into every thing worth knowing, and finding out every thing he possibly could—as he said himself—when he did not understand a thing he hunted up a spirit who could explain it, consequently he is pretty thoroughly posted. And for his accommodating conduct and his uniform courtesy to me, I sincerely hope he will continue to progress at his present velocity until he lands on the very roof of the highest sphere of all, and thus achieves perfection.

BRIEF BIOGRAPHICAL SKETCH OF GEORGE WASHINGTON

THIS day, many years ago precisely, George Washington was born. How full of significance the thought! Especially to those among us who have had a similar experience, though subsequently; and still more especially to the young, who should take him for a model, and faithfully try to be like him, undeterred by the frequency with which the same thing has been attempted by American youths before them and not satisfactorily accomplished. George Washington was the youngest of nine children, eight of whom were the offspring of his uncle and his aunt. As a boy, he gave no promise of the greatness he was one day to achieve. He was ignorant of the commonest accomplishments of youth. He could not even lie. But then he never had any of those precious advantages which are within the reach of the humblest of the boys of the present day. Any boy can lie now. I could lie before I could stand—yet this sort of sprightliness was so common in our family that little notice was taken of it. Young George appears to have had no sagacity whatever. It is related of him that he once chopped down his father's favorite cherry-tree, and then didn't know enough to keep dark about it. He came near going to sea once, as a midshipman; but when his mother represented to him that he must necessarily be absent when he was away from home, and that this must continue to be the case until he got back, the sad truth struck him so forcibly that he ordered his trunk ashore, and quietly but firmly refused to serve in the navy and fight the battles of his king so long as the effect of it would be to discommode his mother. The great rule of his life was, that procrastination was the thief of time, and that we should always do unto others somehow. This is the golden rule. Therefore, he would never discommode his mother.

Young George Washington was actuated in all things by the highest and purest principles of morality, justice, and right. He was a model in every way worthy of the emulation of youth. Young George was always prompt and faithful in the discharge of every duty. It has been said of him, by the historian, that he was always on hand, like a thousand of brick. And well deserved was this compliment. The aggregate of the building material specified might have been largely increased—might have been doubled, even—without doing full justice to these high qualities in the subject of this sketch. Indeed, it would hardly be possible to express in bricks the exceeding promptness and fidelity of young George Washington. His was a soul whose manifold excellencies were beyond the ken and computation of mathematics, and bricks are, at the least, but an inadequate vehicle for the conveyance of a comprehension of the moral sublimity of a nature so pure as his.

Young George W. was a surveyor in early life—a surveyor of an inland port—a sort of county surveyor; and under a commission from Governor Dinwiddie, he set out to survey his way four hundred miles through trackless forests, infested with Indians, to procure the liberation of some English prisoners. The historian says the Indians were the most depraved of their species, and did nothing but lay for white men, whom they killed for the sake of robbing them. Considering that white men only traveled through the country at the rate of one a year, they were probably unable to do what might be termed a land-office business in their line. They did not rob young G. W.; one savage made the attempt, but failed; he fired at the subject of this sketch from behind a tree, but the subject of this sketch immediately snaked him out from behind the tree and took him prisoner.

The long journey failed of success; the French would not give up the prisoners, and Wash went sadly back home again. A regiment was raised to go and make a rescue, and he took command of it. He caught the French out in the rain and tackled them with great intrepidity. He defeated them in ten minutes, and their commander handed in his checks. This was the battle of Great Meadows.

After this, a good while, George Washington became Commander-in-Chief of the American armies, and had an exceedingly dusty time of it all through the Revolution. But every now and then he turned a Jack from the bottom and surprised the enemy. He kept up his lick for seven long years, and hazed the British from Harrisburg to Halifax—and America was free! He served two terms as President, and would have been President yet if he had lived —even so did the people honor the Father of his Country. Let the youth of America take his incomparable character for a model, and try it one jolt, any how. Success is possible —let them remember that—success is possible, though there are chances against it.

I could continue this biography with profit to the rising generation, but I shall have to drop the subject at present, because of other matters which must be attended to.

A TOUCHING STORY OF GEORGE WASHINGTON'S BOYHOOD.

IF it please your neighbor to break the sacred calm of night with the snorting of an unholy trombone, it is your duty to put up with his wretched music and your privilege to pity him for the unhappy instinct that moves him to delight in such discordant sounds. I did not always think thus: this consideration for musical amateurs was born of certain disagreeable personal experiences that once followed the development of a like instinct in myself. Now this infidel over the way, who is learning to play on the trombone, and the slowness of whose progress is almost miraculous, goes on with his harrowing work every night, uncursed by me, but tenderly pitied. Ten years ago, for the same offense, I would have set fire to his house. At that time I was a prey to an amateur violinist for two or three weeks, and the sufferings I endured at his hands are inconceivable. He played "Old Dan Tucker," and he never played any thing else; but he performed that so badly that he could throw me into fits with it if I were awake, or into a nightmare if I were asleep. As long as he confined himself to "Dan Tucker," though, I bore with him and abstained from violence; but when he projected a fresh outrage, and tried to do "Sweet Home," I went over and burnt him out. My next assailant was a wretch who felt a call to play the clarionet. He only played the scale, however, with his distressing instrument, and I let him run the length of his tether, also; but finally, when he branched out into a ghastly tune, I felt my reason deserting me under the exquisite torture, and I sallied forth and burnt him out likewise. During the next two years I burned out an amateur cornet player, a bugler, a bassoon-sophomore, and a barbarian whose talents ran in the base-drum line.

I would certainly have scorched this trombone man if he had moved into my neighborhood in those days. But as I said before, I leave him to his own destruction now, because

I have had experience as an amateur myself, and I feel nothing but compassion for that kind of people. Besides, I have learned that there lies dormant in the souls of all men a penchant for some particular musical instrument, and an unsuspected yearning to learn to play on it, that are bound to wake up and demand attention some day. Therefore, you who rail at such as disturb your slumbers with unsuccessful and demoralizing attempts to subjugate a fiddle, beware! for sooner or later your own time will come. It is customary and popular to curse these amateurs when they wrench you out of a pleasant dream at night with a peculiarly diabolical note; but seeing that we are all made alike, and must all develop a distorted talent for music in the fullness of time, it is not right. I am charitable to my trombone maniac; in a moment of inspiration he fetches a snort, sometimes, that brings me to a sitting posture in bed, broad awake and weltering in a cold perspiration. Perhaps my first thought is, that there has been an earthquake; perhaps I hear the trombone, and my next thought is, that suicide and the silence of the grave would be a happy release from this nightly agony; perhaps the old instinct comes strong upon me to go after my matches; but my first cool, collected thought is, that the trombone man's destiny is upon him, and he is working it out in suffering and tribulation; and I banish from me the unworthy instinct that would prompt me to burn him out.

After a long immunity from the dreadful insanity that moves a man to become a musician in defiance of the will of God that he should confine himself to sawing wood, I finally fell a victim to the instrument they call the accordeon. At this day I hate that contrivance as fervently as any man can, but at the time I speak of I suddenly acquired a disgusting and idolatrous affection for it. I got one of powerful capacity, and learned to play "Auld Lang Syne" on it. It seems to me, now, that I must have been gifted with a sort of inspiration to be enabled, in the state of ignorance in which I then was, to select out of the whole range of musical composition the one solitary tune that sounds vilest and most distressing on the accordeon. I do not suppose there is another tune in the world with which I could have inflicted so much anguish upon my race as I did with that one during my short musical career.

After I had been playing "Lang Syne"

about a week, I had the vanity to think I could improve the original melody, and I set about adding some little flourishes and variations to it, but with rather indifferent success, I suppose, as it brought my landlady into my presence with an expression about her of being opposed to such desperate enterprises. Said she, "Do you know any other tune but that, Mr. Twain?" I told her, meekly, that I did not. "Well, then," said she, "stick to it just as it is; don't put any variations to it, because it's rough enough on the boarders the way it is now."

The fact is, it was something more than simply "rough enough" on them; it was altogether too rough; half of them left, and the other half would have followed, but Mrs. Jones saved them by discharging me from the premises.

I only staid one night at my next lodging-house. Mrs. Smith was after me early in the morning. She said, "You can go, sir; I don't want you here; I have had one of your kind before—a poor lunatic, that played the banjo and danced breakdowns, and jarred the glass all out of the windows. You kept me awake all night, and if you was to do it again, I'd take and mash that thing over your head!" I could see that this woman took no delight in music, and I moved to Mrs. Brown's.

For three nights in succession I gave my new neighbors "Auld Lang Syne," plain and unadulterated, save by a few discords that rather improved the general effect than otherwise. But the very first time I tried the variations the boarders mutinied. I never did find any body that would stand those variations. I was very well satisfied with my efforts in that house, however, and I left it without any regrets; I drove one boarder as mad as a March hare, and another one tried to scalp his mother. I reflected, though, that if I could only have been allowed to give this latter just one more touch of the variations, he would have finished the old woman.

I went to board at Mrs. Murphy's, an Italian lady of many excellent qualities. The very first time I struck up the variations, a haggard, care-worn, cadaverous old man walked into my room and stood beaming upon me a smile of ineffable happiness. Then he placed his hand upon my head, and looking devoutly aloft, he said with feeling unction, and in a voice trembling with emotion, "God bless you, young man! God bless you! for you have done that for me which is beyond all praise. For years I have suffered from an incurable disease, and knowing my doom was sealed and that I must die, I have striven with all my power to resign myself to my fate, but in vain—the love of life was too strong within me. But Heaven bless you, my benefactor! for since I heard you play that tune and those variations, I do not want to live any longer—I am entirely resigned—I am willing to die—in fact, I am anxious to die." And then the old man fell upon my neck and wept a flood of happy tears. I was surprised at these things; but I could not help feeling a little proud at what I had done, nor could I help giving the old gentleman a parting blast in the way of some peculiarly lacerating variations as he went out at the door. They doubled him up like a jack-knife, and the next time he left his bed of pain and suffering he was all right, in a metallic coffin.

My passion for the accordeon finally spent itself and died out, and I was glad when I found myself free from its unwholesome influence. While the fever was upon me, I was a living, breathing calamity wherever I went, and desolation and disaster followed in my wake. I bred discord in families, I crushed the spirits of the light-hearted, I drove the melancholy to despair, I hurried invalids to premature dissolution, and I fear me I disturbed the very dead in their graves. I did incalculable harm, and inflicted untold suffering upon my race with my execrable music; and yet to atone for it all, I did but one single blessed act, in making that weary old man willing to go to his long home.

Still, I derived some little benefit from that accordeon; for while I continued to practice on it, I never had to pay any board — landlords were always willing to compromise, on my leaving before the month was up.

Now, I had two objects in view in writing the foregoing, one of which was to try and reconcile people to those poor unfortunates who feel that they have a genius for music, and who drive their neighbors crazy every night in trying to develop and cultivate it; and the other was to introduce an admirable story about Little George Washington, who could Not Lie, and the Cherry-Tree — or the Apple-Tree — I have forgotten now which, although it was told me only yesterday. And writing such a long and elaborate introductory has caused me to forget the story itself; but it was very touching.

A PAGE FROM A CALIFORNIAN ALMANAC.

AT the instance of several friends who feel a boding anxiety to know beforehand what sort of phenomena we may expect the elements to exhibit during the next month or two, and who have lost all confidence in the various patent medicine almanacs, because of the unaccountable reticence of those works concerning the extraordinary event of the 8th inst., I have compiled the following almanac expressly for the latitude of San Francisco:

Oct. 17.—Weather hazy; atmosphere murky and dense. An expression of profound melancholy will be observable upon most countenances.

Oct. 18.—Slight earthquake. Countenances grow more melancholy.

Oct. 19.—Look out for rain. It would be absurd to look in for it. The general depression of spirits increased.

Oct. 20.—More weather.

Oct. 21.—Same.

Oct. 22.—Light winds, perhaps. If they blow, it will be from the "east'ard, or the nor'ard, or the west'ard, or the suth'ard," or from some general direction approximating more or less to these points of the compass or otherwise. Winds are uncertain — more especially when they blow from whence they cometh and whither they listeth. N. B.—Such is the nature of winds.

Oct. 23.—Mild, balmy earthquakes.

Oct. 24.—Shaky.

Oct. 25.—Occasional shakes, followed by light showers of bricks and plastering. N. B.—Stand from under!

Oct. 26.—Considerable phenomenal atmospheric foolishness. About this time expect more earthquakes; but do not look for them, on account of the bricks.

Oct. 27.—Universal despondency, indicative of approaching disaster. Abstain from smiling, or indulgence in humorous conversation, or exasperating jokes.

Oct. 28.—Misery, dismal forebodings, and

despair. Beware of all light discourse—a joke uttered at this time would produce a popular outbreak.

Oct. 29.—Beware!

Oct. 30.—Keep dark!

Oct. 31.—Go slow!

Nov. 1.—Terrific earthquake. This is the great earthquake month. More stars fall and more worlds are slathered around carelessly and destroyed in November than in any other month of the twelve.

Nov. 2.—Spasmodic but exhilarating earthquakes, accompanied by occasional showers of rain and churches and things.

Nov. 3.—Make your will.

Nov. 4.—Sell out.

Nov. 5.—Select your "last words." Those of John Quincy Adams will do, with the addition of a syllable, thus: "This is the last of earthquakes."

Nov. 6.—Prepare to shed this mortal coil.

Nov. 7.—Shed!

Nov. 8.—The sun will rise as usual, perhaps; but if he does, he will doubtless be staggered some to find nothing but a large round hole eight thousand miles in diameter in the place where he saw this world serenely spinning the day before.

INFORMATION FOR THE MILLION.

A YOUNG man anxious for information writes to a friend residing in Virginia City, Nevada, as follows:

"SPRINGFIELD, Mo., April 12.

"DEAR SIR: My object in writing to you is to have you give me a full history of Nevada. What is the character of its climate? What are the productions of the earth? Is it healthy? What diseases do they die of mostly? Do you think it would be advisable for a man who can make a living in Missouri to emigrate to that part of the country? There are several of us who would emigrate there in the spring if we could ascertain to a certainty that it is a much better country than this. I suppose you know Joel H. Smith? He used to live here; he lives in Nevada now; they say he owns considerable in a mine there. Hoping to hear from you soon, etc., I remain yours, truly,

WILLIAM ——."

The letter was handed in to a newspaper office for reply. For the benefit of all who contemplate moving to Nevada, it is perhaps best to publish the correspondence in its entirety:

DEAREST WILLIAM: Pardon my familiarity —but that name touchingly reminds me of the loved and lost, whose name was similar. I have taken the contract to answer your letter, and although we are now strangers, I feel we shall cease to be so if we ever become acquainted with each other. The thought is worthy of attention, William. I will now respond to your several propositions in the order in which you have fulminated them.

Your object in writing is to have me give you a full history of Nevada. The flattering confidence you repose in me, William, is only equaled by the modesty of your request. I could detail the history of Nevada in five hundred pages octavo; but as you have never done me any harm, I will spare you, though it will be apparent to every body that I would be justified in taking advantage of you if I were a mind to. However, I will condense. Nevada was discovered many years ago by the Mormons, and was called Carson county. It only became Nevada in 1861, by act of Congress. There is a popular tradition that the Almighty created it; but when you come to see it, William, you will think differently. Do not let that discourage you, though. The country looks something like a singed cat, owing to the scarcity of shrubbery, and also resembles that animal in the respect that it has more merits than its personal appearance would seem to indicate. The Grosch brothers found the first silver lead here in 1857. They also founded Silver City, I believe. Signify to your friends, however, that all the mines here do not pay dividends as yet; you may make this statement with the utmost unyielding inflexibility—it will not be contradicted from this quarter. The population of this Territory is about 35,000, one half of which number reside in the united cities of Virginia and Gold Hill. However, I will discontinue this history for the present, lest I get you too deeply interested in this distant land, and cause you to neglect your family or your religion. But I will address you again upon the subject next year. In the mean time, allow me to answer your inquiry as to the character of our climate.

It has no character to speak of, William, and alas! in this respect it resembles many, ah! too many chambermaids in this wretched, wretched world. Sometimes we have the seasons in their regular order, and then again we have winter all the summer, and summer all winter. Consequently, we have never yet come across an almanac that would just exactly fit this latitude. It is mighty regular about not raining, though, William. It will start in here in November and rain about four, and sometimes as much as seven days on a stretch; after that you may loan out your umbrella for twelve months, with the serene confidence which a Christian feels in four aces. Sometimes the winter begins in November and winds up in June; and sometimes there is a bare suspicion of winter in March and April, and summer all the balance of the year. But as a general thing, William, the climate is good, what there is of it.

What are the productions of the earth? You mean in Nevada, of course. On our ranches here any thing can be raised that can be produced on the fertile fields of Missouri. But ranches are very scattering—as scattering, perhaps, as lawyers in heaven. Nevada, for the most part, is a barren waste of sand, embellished with melancholy sage-brush, and fenced in with snow-clad mountains. But these ghastly features were the salvation of the land, William; for no rightly constituted American would have ever come here if the place had been easy of access, and none of our pioneers would have staid after they got here, if they had not felt satisfied that they could not find a smaller chance for making a living anywhere else. Such is man, William, as he crops out in America.

"Is it healthy?" Yes, I think it is as healthy here as it is in any part of the West. But never permit a question of that kind to vegetate in your brain, William; because as long as Providence has an eye on you, you will not be likely to die until your time comes.

"What diseases do they die of mostly?" Well, they used to die of conical balls and cold steel, mostly, but here lately erysipelas and the intoxicating bowl have got the bulge on those things, as was very justly remarked by Mr. Rising last Sunday. I will observe, for your information, William, that Mr. Rising is our Episcopal minister, and has done as much as any man among us to redeem this community from its pristine state of semi-barbarism. We are afflicted with all the diseases incident to the same latitude in the States, I believe, with one or two added and half a dozen subtracted on account of our superior altitude. However, the doctors are about as successful here, both in killing and curing, as they are anywhere.

Now, as to whether it would be advisable for a man who can make a living in Missouri to emigrate to Nevada, I confess I am somewhat mixed. If you are not content in your present condition, it naturally follows that you would be entirely satisfied if you could make either more or less than a living. You would exult in the cheerful exhilaration always produced by a change. Well, you can find your opportunity here, where, if you retain your health, and are sober and industrious, you will inevitably make more than a living, and if you don't, you won't. You can rely upon this statement, William. It contemplates any line of business except the selling of tracts. You can not sell tracts here, William; the people take no interest in tracts; the very best efforts in the tract line—even with pictures on them—have met with no encouragement. Besides, the newspapers have been interfering; a man gets his regular text or so from the Scriptures in his paper, along with the stock sales and the war news, every day now. If you are in the tract business, William, take no chances on Washoe; but you can succeed at any thing else here.

"I suppose you know Joel H. Smith?" Well—the fact is—I believe I don't. Now isn't that singular? Isn't it very singular? And he owns "considerable" in a mine here too. Happy man! Actually owns in a mine here in Nevada Territory, and I never even heard of him. Strange—strange—do you know, William, it is the strangest thing that ever happened to me? And then he not only owns in a mine, but owns "considerable;" that is the strangest part about it—how a man could own considerable in a mine in Washoe, and I not know any thing about it. He is a lucky dog, though. But I strongly suspect that you have made a mistake in the name; I am confident you have; you mean John Smith—I know you do; I know it from the fact that he owns considerable in a mine here, because I sold him the property at a ruinous sacrifice on the very day he arrived here from over the plains. That man will be rich one of these days. I am just as well satisfied of it as I am of any precisely similar instance of the kind that has come under my notice. I said as much to him yesterday, and he said he was satisfied of it also. But he did not say it with that air of triumphant exultation which a heart like mine so delights to behold in one to whom I have endeavored to be a benefactor in a small way. He looked pensive awhile, but, finally, says he, "Do you know, I think I'd a been a rich man long ago if they'd ever found the d—d ledge?" That was my idea about it. I always thought, and I still think, that if they ever do find that ledge, his chances will be better than they are now. I guess Smith will be all right one of these centuries, if he keeps up his assessments—he is a young man yet. Now, William, I have taken a liking to you, and I would like to sell you "considerable" in a mine in Washoe. Let me hear from you on the subject. Greenbacks at par is as good a thing as I want. But seriously, William, don't you ever invest in a mining stock which you don't know any thing about; beware of John Smith's experience!

You hope to hear from me soon? Very good. I shall also hope to hear from you soon, about that little matter above referred to. Now, William, ponder this epistle well; never mind the sarcasm here and there, and the nonsense, but reflect upon the plain facts set forth, because they *are* facts, and are meant to be so understood and believed.

Remember me affectionately to your friends and relations, and especially to your venerable grandmother, with whom I have not the pleasure to be acquainted—but that is of no consequence, you know. I have been in your town many a time, and all the towns of the neighboring counties—the hotel-keepers will recollect me vividly. Remember me to them—I bear them no animosity.

Yours affectionately.

THE LAUNCH OF THE STEAMER CAPITAL.

I GET MR. MUFF NICKERSON TO GO WITH ME AND ASSIST IN REPORTING THE GREAT STEAMBOAT LAUNCH.—HE RELATES THE INTERESTING HISTORY OF THE TRAVELING PANORAMIST.

I WAS just starting off to see the launch of the great steamboat Capital, on Saturday week, when I came across Mulph, Mulff, Muff, Mumph, Murph, Mumf, Murf, Mumford, Mulford, Murphy Nickerson—(he is well known to the public by all these names, and I can not say which is the right one)—bound on the same errand.

This was the man I wanted.

We set out in a steamer whose decks were crowded with persons of all ages, who were happy in their nervous anxiety to behold the novelty of a steamboat launch.

As we approached the spot where the launch was to take place, a gentleman from Reese River, by the name of Thompson, came up, with several friends, and said he had been prospecting on the main deck, and had found an object of interest—a bar. This was all very well, and showed him to be a man of parts; but like many another man who produces a favorable impression by an introductory remark replete with wisdom, he followed it up with a vain and unnecessary question—Would we take a drink? This to me!—This to M. M. M., etc., Nickerson!

We proceeded, two by two, arm-in-arm, down to the bar in the nether regions, chatting pleasantly and elbowing the restless multitude. We took pure, cold, health-giving water, with some other things in it, and clinked our glasses together, and were about to drink, when Smith, of Excelsior, drew forth his handkerchief and wiped away a tear; and then, noticing that the action had excited some attention, he explained it by recounting a most affecting incident in the history of a venerated aunt of his—now deceased—and said that, although long years had passed since the touching event he had narrated, he could never take a drink without thinking of the kind-hearted old lady.

Mr. Nickerson blew his nose, and said with deep emotion that it gave him a better opinion of human nature to see a man who had had a good aunt, eternally and forever thinking about her.

This episode reminded Jones, of Mud Springs, of a circumstance which happened many years ago in the home of his childhood, and we held our glasses untouched and rested our elbows on the counter, while we listened with rapt attention to his story.

There was something in it about a good-natured, stupid man, and this reminded Thompson, of Reese River, of a person of the same kind whom he had once fallen in with while traveling through the back settlements of one of the Atlantic States, and we postponed drinking until he should give us the facts in the case. The hero of the tale had unintentionally created some consternation at a camp-meeting by one of his innocent asinine freaks; and this reminded Mr. M. Nickerson of a reminiscence of his temporary sojourn in the interior of Connecticut some months ago; and again our uplifted glasses were staid on their way to our lips, and we listened attentively to

THE ENTERTAINING HISTORY OF THE SCRIPTURAL PANORAMIST.

[I give the history in Mr. Nickerson s own language.]

There was a fellow traveling around, in that country, (said Mr. Nickerson,) with a moral religious show—a sort of a scriptural panorama—and he hired a wooden-headed old slab to play the piano for him. After the first night's performance, the showman says:

"My friend, you seem to know pretty much all the tunes there are, and you worry along first-rate. But then didn't you notice that sometimes last night the piece you happened to be playing was a little rough on the proprieties, so to speak—didn't seem to jibe with the general gait of the picture that was passing at the time, as it were—was a little foreign to the subject, you know—as if you didn't either trump or follow suit, you understand?"

"Well, no," the fellow said; he hadn't noticed, but it might be; he had played along just as it came handy.

So they put it up that the simple old dummy was to keep his eye on the panorama after that, and as soon as a stunning picture was reeled out, he was to fit it to a dot with a piece of music that would help the audience get the idea of the subject, and warm them up like a camp-meeting revival. That sort of thing would corral their sympathies, the showman said.

There was a big audience that night—mostly middle-aged and old people who belonged to the church and took a strong interest in Bible matters, and the balance were pretty much young bucks and heifers—*they* always come out strong on panoramas, you know, because it gives them a chance to taste one another's mugs in the dark.

Well, the showman began to swell himself up for his lecture, and the old mud-dobber tackled the piano and run his fingers up and down once or twice to see that she was all right, and the fellows behind the curtain commenced to grind out the panorama. The showman balanced his weight on his right foot, and propped his hands on his hips, and flung his eye over his shoulder at the scenery, and says:

"Ladies and gentlemen, the painting now before you illustrates the beautiful and touching parable of the Prodigal Son. Observe the happy expression just breaking over the features of the poor suffering youth—so worn and weary with his long march; note also the ecstasy beaming from the uplifted countenance of the aged father, and the joy that sparkles in the eyes of the excited group of youths and maidens, and seems ready to burst in a welcoming chorus from their lips. The lesson, my friends, is as solemn and instructive as the story is tender and beautiful."

The mud-dobber was all ready, and the second the speech was finished he struck up:

> "Oh! we'll all get blind drunk
> When Johnny comes marching home!"

Some of the people giggled, and some groaned a little. The showman couldn't say a word. He looked at the piano-sharp; but he was all lovely and serene—*he* didn't know there was any thing out of gear.

The panorama moved on, and the showman drummed up his grit and started in fresh:

"Ladies and gentlemen, the fine picture now

unfolding itself to your gaze exhibits one of the most notable events in Bible history—our Saviour and his disciples upon the Sea of Galilee. How grand, how awe-inspiring are the reflections which the subject invokes! What sublimity of faith is revealed to us in this lesson from the sacred writings! The Saviour rebukes the angry waves, and walks securely upon the bosom of the deep!"

All around the house they were whispering, "Oh! how lovely! how beautiful!" and the orchestra let himself out again:

> "Oh! a life on the ocean wave,
> And a home on the rolling deep!"

There was a good deal of honest snickering turned on this time, and considerable groaning, and one or two old deacons got up and went out. The showman gritted his teeth and cursed the piano man to himself; but the fellow sat there like a knot on a log, and seemed to think he was doing first-rate.

After things got quiet, the showman thought he would make one more stagger at it, any how, though his confidence was beginning to get mighty shaky. The supes started the panorama to grinding along again, and he says:

"Ladies and gentlemen, this exquisite painting illustrates the raising of Lazarus from the dead by our Saviour. The subject has been handled with rare ability by the artist, and such touching sweetness and tenderness of expression has he thrown into it, that I have known peculiarly sensitive persons to be even affected to tears by looking at it. Observe the half-confused, half-inquiring look, upon the countenance of the awakening Lazarus. Observe, also, the attitude and expression of the Saviour, who takes him gently by the sleeve of his shroud with one hand, while he points with the other toward the distant city."

Before any body could get off an opinion in the case, the innocent old ass at the piano struck up:

> "Come, rise up, William Ri-i-ley
> And go along with me!"

It was rough on the audience, you bet you. All the solemn old flats got up in a huff to go, and every body else laughed till the windows rattled.

The showman went down and grabbed the orchestra, and shook him up, and says:

"That lets you out, you know, you chowder-headed old clam! Go to the doorkeeper and get your money, and cut your stick! vamose the ranche! Ladies and gentlemen, circumstances over which I have no control compel me prematurely to dismiss——"

"By George! it was splendid! Come! all hands! let's take a drink!"

It was Phelim O'Flannigan, of San Luis Obispo, who interrupted. I had not seen him before.

"What was splendid?" I inquired.

"The launch!"

Our party clinked glasses once more, and drank in respectful silence.

P. S.—You will excuse me from making a model report of the great launch. I was with Mulf Nickerson, who was going to "explain the whole thing to me as clear as glass;" but, you see, they launched the boat with such indecent haste, that we never got a chance to see it. It was a great pity, because Mulph Nickerson understands launches as well as any man.

ORIGIN OF ILLUSTRIOUS MEN.

JOHN SMITH was the son of his father. He formerly lived in New-York and other places, but he has removed to San Francisco now.

William Smith was the son of his mother. This party's grandmother is deceased. She was a brick.

John Brown was the son of old Brown. The body of the latter lies mouldering in the grave.

Edward Brown was the son of old Brown by a particular friend.

Henry Jones was the son of a sea-cook.

Ed Jones was a son of a gun.

John Jones was a son of temperance.

In early life Gabriel Jones was actually a shoemaker. He is a shoemaker yet.

Previous to the age of eighty-five, Caleb

Jones had never given evidence of extraordinary ability. He has never given any since.

Patrick Murphy is said to have been of Irish extraction.

James Peterson was the son of a common weaver, who was so miraculously poor that his friends were encouraged to believe that in case the Scriptures were carried out he would "inherit the earth." He never got his property.

John Davis's father was the son of a soap-boiler, and not a very good soap-boiler at that. John never arrived at maturity—died in childbirth—he and his mother.

John Johnson was a blacksmith. He died. It was published in the papers, with a head over it, "Deaths." It was, therefore, thought he died to gain notoriety. He has got an aunt living somewhere.

Up to the age of thirty-four Hosea Wilkerson never had any home but Home Sweet Home, and even then he had it to sing himself. At one time it was believed that he would have been famous if he became celebrated. He died. He was greatly esteemed for his many virtues. There was not a dry eye in the crowd when they planted him.

ADVICE FOR GOOD LITTLE GIRLS.

GOOD little girls ought not to make mouths at their teachers for every trifling offense. This kind of retaliation should only be resorted to under peculiarly aggravating circumstances.

If you have nothing but a rag doll stuffed with saw-dust, while one of your more fortunate little playmates has a costly china one, you should treat her with a show of kindness, nevertheless. And you ought not to attempt to make a forcible swap with her unless your conscience would justify you in it, and you know you are able to do it.

You ought never to take your little brother's "chawing-gum" away from him by main force; it is better to rope him in with the promise of the first two dollars and a half you find floating down the river on a grindstone. In the artless simplicity natural to his time of life, he will regard it as a perfectly fair transaction. In all ages of the world this eminently plausible fiction has lured the obtuse infant to financial ruin and disaster.

If at any time you find it necessary to correct your brother, do not correct him with mud—never on any account throw mud at him, because it will soil his clothes. It is better to scald him a little; for then you attain two desirable results—you secure his immediate attention to the lesson you are inculcating, and, at the same time, your hot water will have a tendency to remove impurities from his person—and possibly the skin also, in spots.

If your mother tells you to do a thing, it is wrong to reply that you won't. It is better and more becoming to intimate that you will do as she bids you, and then afterward act quietly in the matter according to the dictates of your better judgment.

You should ever bear in mind that it is to your kind parents that you are indebted for your food and your nice bed and your beautiful clothes, and for the privilege of staying home from school when you let on that you are sick. Therefore you ought to respect their little prejudices and humor their little whims and put up with their little foibles, until they get to crowding you too much.

Good little girls should always show marked deference for the aged. You ought never to "sass" old people—unless they "sass" you first.

CONCERNING CHAMBERMAIDS.

AGAINST all chambermaids, of whatsoever age or nationality, I launch the curse of bachelordom! Because:

They always put the pillows at the opposite end of the bed from the gas-burner, so that while you read and smoke before sleeping, (as is the ancient and honored custom of bachelors,) you have to hold your book aloft, in an uncomfortable position, to keep the light from dazzling your eyes.

When they find the pillows removed to the other end of the bed in the morning, they receive not the suggestion in a friendly spirit; but, glorying in their absolute sovereignty, and unpitying your helplessness, they make the bed just as it was originally, and gloat in secret over the pang their tyranny will cause you.

Always after that, when they find you have transposed the pillows, they undo your work, and thus defy and seek to embitter the life that God has given you.

If they can not get the light in an inconvenient position any other way, they move the bed.

If you pull your trunk out six inches from the wall, so that the lid will stay up when you open it, they always shove that trunk back again. They do it on purpose.

If you want the spittoon in a certain spot, where it will be handy, they don't, and so they move it.

They always put your other boots into inaccessible places. They chiefly enjoy depositing them as far under the bed as the wall will permit. It is because this compels you to get down in an undignified attitude and make wild sweeps for them in the dark with the boot-jack, and swear.

They always put the match-box in some other place. They hunt up a new place for it every day, and put up a bottle, or other perishable glass thing, where the box stood before. This is to cause you to break that glass thing, groping in the dark, and get yourself into trouble.

They are forever and ever moving the furniture. When you come in, in the night, you can calculate on finding the bureau where the wardrobe was in the morning. And when you go out in the morning, if you leave the slop-bucket by the door and rocking-chair by the window, when you come in at midnight, or thereabouts, you will fall over that rocking-chair, and you will proceed toward the window and sit down in that slop-tub. This will disgust you. They like that.

No matter where you put any thing, they are not going to let it stay there. They will take it and move it the first chance they get. It is their nature. And, besides, it gives them pleasure to be mean and contrary this way. They would die if they couldn't be villains.

They always save up all the old scraps of printed rubbish you throw on the floor, and stack them up carefully on the table, and start the fire with your valuable manuscripts. If there is any one particular old scrap that you are more down on than any other, and which you are gradually wearing your life out trying to get rid of, you may take all the pains you possibly can in that direction, but it won't be of any use, because they will always fetch that old scrap back and put it in the same old place again every time. It does them good.

And they use up more hair-oil than any six men. If charged with purloining the same, they lie about it. What do they care about a hereafter? Absolutely nothing.

If you leave your key in the door for convenience sake, they will carry it down to the office and give it to the clerk. They do this under the vile pretense of trying to protect your property from thieves; but actually they do it because they want to make you tramp back down-stairs after it when you come home tired, or put you to the trouble of sending a waiter for it, which waiter will expect you to pay him something. In which case I suppose the degraded creatures divide.

They keep always trying to make your bed before you get up, thus destroying your rest and inflicting agony upon you; but after you get up, they don't come any more till next day.

They do all the mean things they can think of, and they do them just out of pure cussedness, and nothing else.

Chambermaids are dead to every human instinct.

I have cursed them in behalf of outraged bachelordom. They deserve it. If I can get a bill through the Legislature abolishing chambermaids, I mean to do it.

REMARKABLE INSTANCES OF PRESENCE OF MIND.

THE steamer Ajax encountered a terrible storm on her down trip from San Francisco to the Sandwich Islands. It tore her light spars and rigging all to shreds and splinters, upset all furniture that could be upset, and spilled passengers around and knocked them hither and thither with a perfect looseness. For forty-eight hours no table could be set, and every body had to eat as best they might under the circumstances. Most of the party went hungry, though, and attended to their praying. But there was one set of "seven-up" players who nailed a card-table to the floor and stuck to their game through thick and thin. Captain F——, of a great banking-house in San Francisco, a man of great coolness and presence of mind, was of this party. One night the storm suddenly culminated in a climax of unparalleled fury; the vessel went down on her beam ends, and every thing let go with a crash—passengers, tables, cards, bottles—every thing came clattering to the floor in a chaos of disorder and confusion. In a moment fifty sore distressed and pleading voices ejaculated, "O Heaven! help us in our extremity!" and one voice rang out clear and sharp above the plaintive chorus and said, "Remember, boys, I played the tray for low!" It was one of the gentlemen I have mentioned who spoke. And the remark showed good presence of mind and an eye to business.

Lewis L——, of a great hotel in San Francisco, was a passenger. There were some savage grizzly bears chained in cages on deck. One night, in the midst of a hurricane, which was accompanied by rain and thunder and lightning, Mr. L. came up, on his way to bed. Just as he stepped into the pitchy darkness of the deck and reeled to the still more pitchy motion of the vessel, (bad,) the captain sang out hoarsely through his speaking-trumpet, "Bear a hand aft, there!" The words were sadly marred and jumbled by the roaring wind. Mr. L—— thought the captain said, "The bears are after you there!" and he "let go all holts" and went down into his boots. He murmured, "I knew how it was going to be—I just knew it from the start—I said all along that those bears would get loose some time; and now I'll be the first man that they'll snatch. Captain! captain!—can't hear me—storm roars so! O God! what a fate! I have avoided wild beasts all my life, and now to be eaten by a grizzly bear in the middle of the ocean, a thousand miles from land! Captain! O captain!—bless my soul, there's one of them—I've got to cut and run!" And he did cut and run, and smashed through the door of the first state-room he came to. A gentleman and his wife were in it. The gentleman exclaimed, "Who's that?" The refugee gasped out, "O great Scotland! those bears are loose, and just raising merry hell all over the ship!" and then sank down exhausted. The gentleman sprang out of bed and locked the door, and prepared for a siege. After a while, no assault being made, a reconnoissance was made from the window, and a vivid flash of lightning revealed a clear deck. Mr. L—— then made a dart for his own state-room, gained it, locked himself in, and felt that his body's salvation was accomplished, and by little less than a miracle. The next day the subject of this memoir, though still very feeble and nervous, had the hardihood to make a joke upon his adventure. He said that when he found himself in so tight a place (as he thought) he didn't bear it with much fortitude, and when he found himself safe at last in his state-room, he regarded it as the bearest escape he had ever had in his life. He then went to bed, and did not get up again for nine days. This unquestionably bad joke cast a gloom over the whole ship's company, and no effort was sufficient to restore their wonted cheerfulness until the vessel reached her port, and other scenes erased it from their memories.

HONORED AS A CURIOSITY IN HONOLULU.

IF you get into conversation with a stranger in Honolulu, and experience that natural desire to know what sort of ground you are treading on by finding out what manner of man your stranger is, strike out boldly and address him as "Captain." Watch him narrowly, and if you see by his countenance that you are on the wrong track, ask him where he preaches. It is a safe bet that he is either a missionary or captain of a whaler. I became personally acquainted with seventy-two captains and ninety-six missionaries. The captains and ministers form one half of the population; the third fourth is composed of common Kanakas and mercantile foreigners and their families; and the final fourth is made up of high officers of the Hawaiian government. And there are just about cats enough for three apiece all around.

A solemn stranger met me in the suburbs one day, and said:

"Good morning, your reverence. Preach in the stone church yonder, no doubt?"

"No, I don't. I'm not a preacher."

"Really, I beg your pardon, captain. I trust you had a good season. How much oil——"

"Oil! Why, what do you take me for? I'm not a whaler."

"Oh! I beg a thousand pardons, your Excellency. Major-General in the household troops, no doubt? Minister of the Interior, likely? Secretary of War? First Gentleman of the Bed-chamber? Commissioner of the Royal——"

"Stuff! man. I'm no official. I'm not connected in any way with the government."

"Bless my life! Then who the mischief are you? what the mischief are you? and how the mischief did you get here? and where in thunder did you come from?"

"I'm only a private personage—an unassuming stranger—lately arrived from America."

"No! Not a missionary! not a whaler!

not a member of his Majesty's government! not even Secretary of the Navy! Ah! heaven! it is too blissful to be true; alas! I do but dream. And yet that noble, honest countenance—those oblique, ingenuous eyes—that massive head, incapable of—of—any thing; your hand; give me your hand, bright waif. Excuse these tears. For sixteen weary years I have yearned for a moment like this, and——"

Here his feelings were too much for him, and he swooned away. I pitied this poor creature from the bottom of my heart. I was deeply moved. I shed a few tears on him, and kissed him for his mother. I then took what small change he had, and "shoved."

THE STEED "OAHU."

THE landlord of the American hotel at Honolulu said the party had been gone nearly an hour, but that he could give me my choice of several horses that could easily overtake them. I said, Never mind—I preferred a safe horse to a fast one—I would like to have an excessively gentle horse—a horse with no spirit whatever—a lame one, if he had such a thing. Inside of five minutes I was mounted, and perfectly satisfied with my outfit. I had no time to label him, "This is a horse," and so if the public took him for a sheep I can not help it. I was satisfied, and that was the main thing. I could see that he had as many fine points as any man's horse, and I just hung my hat on one of them, behind the saddle, and swabbed the perspiration from my face and started. I named him after this island, "Oahu," (pronounced O-waw-hoo.)

The first gate he came to he started in; I had neither whip nor spur, and so I simply argued the case with him. He firmly resisted argument, but ultimately yielded to insult and abuse. He backed out of that gate and steered for another one on the other side of the street. I triumphed by my former process. Within the next six hundred yards he crossed the street fourteen times, and attempted thirteen gates, and in the mean time the tropical sun was beating down and threatening to cave the top of my head in, and I was literally dripping with perspiration and profanity. (I am only human, and I was sorely aggravated; I shall behave better next time.) He quit the gate business after that, and went along peaceably enough, but absorbed in meditation. I noticed this latter circumstance, and it soon began to fill me with the gravest apprehension. I said to myself, This malignant brute is planning some new outrage—some fresh deviltry or other; no horse ever thought over a subject so profoundly as this one is doing just for nothing. The more this thing preyed upon my mind the more uneasy I became, until at last the suspense became unbearable, and I dismounted to see if there was any thing wild in his eye; for I had heard that the eye of this noblest of our domestic animals is very expressive. I can not describe what a load of anxiety was lifted from my mind when I found that he was only asleep. I woke him up and started him into a faster walk, and then the inborn villainy of his nature came out again. He tried to climb over a stone wall five or six feet high. I saw that I must apply force to this horse, and that I might as well begin first as last. I plucked a stout switch from a tamarind tree, and the moment he saw it he gave in. He broke into a convulsive sort of a canter, which had three short steps in it and one long one, and reminded me alternately of the clattering shake of the great earthquake and the sweeping plunging of the Ajax in a storm.

A STRANGE DREAM.

DREAMED AT THE VOLCANO HOUSE, CRATER OF "KILAUEA," SANDWICH ISLANDS, APRIL 1, 1866.

ALL day long I have sat apart and pondered over the mysterious occurrences of last night. . . There is no link lacking in the chain of incidents — my memory presents each in its proper order with perfect distinctness, but still——

However, never mind these reflections—I will drop them and proceed to make a simple statement of the facts.

Toward eleven o'clock, it was suggested that the character of the night was peculiarly suited to viewing the mightiest active volcano on the earth's surface in its most impressive sublimity. There was no light of moon or star in the inky heavens to mar the effect of the crater's gorgeous pyrotechnics.

In due time I stood, with my companion, on the wall of the vast cauldron which the natives, ages ago, named *Hale mau mau*—the abyss wherein they were wont to throw the remains of their chiefs, to the end that vulgar feet might never tread above them. We stood there, at dead of night, a mile above the level of the sea, and looked down a thousand feet upon a boiling, surging, roaring ocean of fire!—shaded our eyes from the blinding glare, and gazed far away over the crimson waves with a vague notion that a supernatural fleet, manned by demons and freighted with the damned, might presently sail up out of the remote distance; started when tremendous thunder-bursts shook the earth, and followed with fascinated eyes the grand jets of molten lava that sprang high up toward the zenith and exploded in a world of fiery spray that lit up the sombre heavens with an infernal splendor.

"What is your little bonfire of Vesuvius to this?"

My ejaculation roused my companion from his reverie, and we fell into a conversation appropriate to the occasion and the surroundings. We came at last to speak of the ancient custom of casting the bodies of dead chieftains into this fearful caldron; and my comrade, who is of the blood royal, mentioned that the founder of his race, old King Kamehameha the First—that invincible old pagan Alexander—had found other sepulture than the burning depths of the *Hale mau mau*. I grew interested at once; I knew that the mystery of what became of the corpse of the warrior king had never been fathomed; I was aware that there was a legend connected with this matter; and I felt as if there could be no more fitting time to listen to it than the present. The descendant of the Kamehamehas said:

"The dead king was brought in royal state down the long, winding road that descends from the rim of the crater to the scorched and chasm-riven plain that lies between the *Hale mau mau* and those beetling walls yonder in the distance. The guards were set and the troops of mourners began the weird wail for the departed. In the middle of the night came a sound of innumerable voices in the air, and the rush of invisible wings; the funeral torches wavered, burned blue, and went out. The mourners and watchers fell to the ground paralyzed by fright, and many minutes elapsed before any one dared to move or speak; for they believed that the phantom messengers of the dread Goddess of Fire had been in their midst. When at last a torch was lighted, the bier was vacant — the dead monarch had been spirited away! Consternation seized upon all, and they fled out of the crater. When day dawned, the multitude returned and began the search for the corpse. But not a footprint, not a sign was ever found. Day after day the search was continued, and every cave in the great walls, and every chasm in the plain, for miles around, was examined, but all to no purpose; and from that day to this the resting-place of the lion king's bones is an unsolved mystery. But years afterward, when the grim prophetess Wiahowakawak lay on her deathbed, the Goddess *Pele* appeared to her in a vision, and told her that eventually the secret would be revealed, and in a remarkable manner, but not until the great *Kauhuhu*, the Shark God, should desert the sacred cavern *Aua Puhi*, in the Island of Molokai, and the waters of the sea should no more visit it, and its floors should become dry. Ever since that time the simple, confiding natives have watched for the sign. And now, after many and many a summer has come and gone, and they who were in the flower of youth then have waxed old and died, the day is at hand! The great Shark God has deserted the *Aua Puhi:* a month ago, for the first time within the records of the ancient legends, the waters of the sea ceased to flow into the cavern, and its stony pavement is become dry! As you may easily believe, the news of this event spread like wildfire through the islands, and now the natives are looking every hour for the miracle which is to unvail the mystery and reveal the secret grave of the dead hero."

After I had gone to bed I got to thinking of the volcanic magnificence we had witnessed, and could not go to sleep. I hunted up a book and concluded to pass the time in reading. The first chapter I came upon related several instances of remarkable revelations, made to men through the agency of dreams—of roads and houses, trees, fences, and all manner of landmarks, shown in visions and recognized afterward in waking hours, and which served to point the way to some dark mystery or other.

At length I fell asleep, and dreamed that I was abroad in the great plain that skirts the *Hale mau mau.* I stood in a sort of twilight which softened the outlines of surrounding objects, but still left them tolerably distinct. A gaunt, muffled figure stepped out from the shadow of a rude column of lava, and moved away with a slow and measured step, beckoning me to follow. I did so. I marched down, down, down, hundreds of feet, upon a narrow trail which wound its tortuous course through piles and pyramids of seamed and blackened lava, and under overhanging masses of sulphur formed by the artist hand of nature into an infinitude of fanciful shapes. The thought crossed my mind that possibly my phantom guide might lead me down among the bowels of the crater, and then disappear and leave me to grope my way through its mazes, and work out my deliverance as best I might; and so, with an eye to such a contingency, I picked up a stone, and "blazed" my course by breaking off a projecting corner, occasionally, from lava walls and festoons of sulphur. Finally we turned into a cleft in the crater's side, and pursued our way through its intricate windings for many a fathom down toward the home of the subterranean fires, our course lighted all the while by a ruddy glow which filtered up through innumerable cracks and crevices, and which afforded me occasional glimpses of the flood of molten fire boiling and hissing in the profound depths beneath us. The heat was intense, and the sulphurous atmosphere suffocating; but I toiled on in the footsteps of my stately guide, and uttered no complaint. At last we came to a sort of rugged chamber whose sombre and blistered walls spake with mute eloquence of some fiery tempest that had spent its fury here in a bygone age. The spectre pointed to a great boulder at the farther extremity—stood and pointed, silent and motionless, for a few fleeting moments, and then disappeared! "The grave of the last Kamehameha!" The words swept mournfully by, from unknown source, and died away in the distant corridors of my prison-house, and I was alone in the bowels of the earth, in the home of desolation, in the presence of death!

My first frightened impulse was to fly, but a

stronger impulse arrested me and impelled me to approach the massive boulder the spectre had pointed at. With hesitating step I went forward and stood beside it—nothing there. I grew bolder, and walked around and about it, peering shrewdly into the shadowy half-light that surrounded it—still nothing. I paused to consider what to do next. While I stood irresolute, I chanced to brush the ponderous stone with my elbow, and lo! it vibrated to my touch! I would as soon have thought of starting a kiln of bricks with my feeble hand. My curiosity was excited. I bore against the boulder, and it still yielded; I gave a sudden push with my whole strength, and it toppled from its foundation with a crash that sent the echoes thundering down the avenues and passages of the dismal cavern! And there, in a shallow excavation over which it had rested, lay the crumbling skeleton of King Kamehameha the Great, thus sepulchred in long years, by supernatural hands! The bones could be none other; for with them lay the rare and priceless crown of *pulamalama* coral, sacred to royalty, and *tabu* to all else beside. A hollow human groan issued out of the——

I woke up. How glad I was to know it was all a dream! "This comes of listening to the legend of the noble lord—of reading of those lying dream revelations—of allowing myself to be carried away by the wild beauty of old *Kileana* at midnight—of gorging too much pork and beans for supper!" And so I turned over and fell asleep again. And dreamed the same dream precisely as before; followed the phantom—"blazed" my course—arrived at the grim chamber—heard the sad spirit voice—overturned the massy stone—beheld the regal crown and the decaying bones of the great king!

I woke up, and reflected long upon the curious and singularly vivid dream, and finally muttered to myself, "This—this is becoming serious!"

I fell asleep again, and again I dreamed the same dream, without a single variation! I slept no more, but tossed restlessly in bed and longed for daylight. And when it came, I wandered forth, and descended to the wide plain in the crater. I said to myself, "I am not superstitious; but if there is any thing in that dying woman's prophecy, I am the instrument appointed to uncurtain this ancient mystery." As I walked along, I even half expected to see my solemn guide step out from some nook in the lofty wall, and beckon me to come on. At last when I reached the place where I had first seen him in my dream, I recognized every surrounding object, and there, winding down among the blocks and fragments of lava, saw the very trail I had traversed in my vision! I resolved to traverse it again, come what might. I wondered if, in my unreal journey, I had "blazed" my way, so that it would stand the test of stern reality; and thus wondering, a chill went to my heart when I came to the first stony projection I had broken off in my dream, and saw the fresh new fracture, and the dismembered fragment lying on the ground! My curiosity rose up and banished all fear, and I hurried along as fast as the rugged road would allow me. I looked for my other "blazes," and found them; found the cleft in the wall; recognized all its turnings; walked in the light that ascended from the glowing furnaces visible far below; sweated in the close, hot atmosphere, and breathed the sulphurous smoke—and at last I stood hundreds of feet beneath the peaks of *Kileana* in the ruined chamber, and in the presence of the mysterious boulder!

"This is no dream," I said; "this is a revelation from the realm of the supernatural; and it becomes not me to longer reason, conjecture, suspect, but blindly to obey the impulses given me by the unseen power that guides me."

I moved with a slow and reverent step toward the stone and bore against it. It yielded perceptibly to the pressure. I brought my full weight and strength to bear, and surged against it. It yielded again; but I was so enfeebled by my toilsome journey that I could not overthrow it. I rested a little, and then raised an edge of the boulder by a strong, steady push, and placed a small stone under it, to keep it from sinking back to its place. I rested again, and then repeated the process. Before long, I had added a third prop, and had got the edge of the boulder considerably elevated. The labor and the close atmosphere together were so exhausting, however, that I was obliged to lie down then, and recuperate my strength by a longer season of rest. And so, hour after hour I labored, growing more and more weary, but still upheld by a fascination which I felt was infused into me by the invisible powers whose will I was working. At last I concentrated my strength in a final effort, and the stone rolled from its position.

I can never forget the overpowering sense of awe that sank down like a great darkness upon my spirit at that moment. After a solemn pause to prepare myself, with bowed form and uncovered head, I slowly turned my gaze till it rested upon the spot where the great stone had lain.

There wasn't any bones there!

. . . .

I just said to myself, "Well, if this an't the blastedest, infernalest swindle that ever I've come across yet, I wish I may never!"

And then I scratched out of there, and marched up here to the Volcano House, and got out my old raw-boned fool of a horse, "Oahu," and "lammed" him till he couldn't stand up without leaning against something.

You can not bet any thing on dreams.

SHORT AND SINGULAR RATIONS.

AS many will remember, the clipper-ship Hornet, of New-York, was burned at sea on her passage to San Francisco. The disaster occurred in lat. 2° 20′ north, long. 112° 8′ west. After being forty-three days adrift on the broad Pacific, in open boats, the crew and passengers succeeded in making Hawaii. A tribute to the courage and brave endurance of these men has been paid in a letter detailing their sufferings, (the particulars being gathered from their own lips,) from which the following excerpt is made:

On Monday, the thirty-eighth day after the disaster, "we had nothing left," said the third mate, "but a pound and a half of ham—the bone was a good deal the heaviest part of it—and one soup-and-bully tin." These things were divided among the fifteen men, and they ate it all—two ounces of food to each man. I do not count the ham-bone, as that was saved for next day. For some time, now, the poor wretches had been cutting their old boots into small pieces and eating them. They would also pound wet rags to a sort of pulp and eat them.

On the thirty-ninth day the ham-bone was divided up into rations, and scraped with knives and eaten. I said, "You say the two sick men remained sick all through, and after a while two or three had to be relieved from standing watch; how did you get along without medicines?"

The reply was, "Oh! we couldn't have kept them if we'd had them; if we'd had boxes of pills, or any thing like that, we'd have eaten them. It was just as well—we couldn't have kept them, and we couldn't have given them to the sick men alone—we'd have shared them around all alike, I guess." It was said rather in jest, but it was a pretty true jest, no doubt.

After apportioning the ham-bone, the captain cut the canvas cover that had been around the ham into fifteen equal pieces, and each man took his portion. This was the last division of food the captain made. The men broke up the small oaken butter tub, and divided the staves among themselves, and gnawed them up. The shell of a little green turtle was scraped with knives, and eaten to the last shaving. The third mate chewed pieces of boots, and spit them out, but ate nothing except the soft straps of two pairs of boots — ate three on the thirty-ninth day, and saved one for the fortieth.

The men seem to have thought in their own minds of the shipwrecked mariner's last dreadful resort—cannibalism; but they do not appear to have conversed about it. They only thought of the casting lots and killing one of their number as a possibility; but even when they were eating rags, and bone, and boots, and shell, and hard oak wood, they seem to have still had a notion that it was remote. They felt that some one of the company must die soon — which one they well knew; and during the last three or four days of their terrible voyage they were patiently but hungrily waiting for him. I wonder if the subject of these anticipations knew what they were thinking of? He must have known it—he must have felt it. They had even calculated how long he would last. They said to themselves, but not to each other—I think they said, "He will die Saturday—and then!"

There was one exception to the spirit of delicacy I have mentioned—a Frenchman—who kept an eye of strong personal interest upon the sinking man, and noted his failing strength with untiring care and some degree of cheerfulness. He frequently said to Thomas, "I think he will go off pretty soon now, sir; and then we'll eat him!" This is very sad.

Thomas, and also several of the men, state that the sick "Portyghee," during the five days that they were entirely out of provisions, actually ate two silk handkerchiefs and a couple of cotton shirts, besides his share of the boots, and bones, and lumber.

Captain Mitchell was fifty-six years old on the twelfth of June—the fortieth day after the burning of the ship and the third day before the boat's crew reached land. He said it looked somewhat as if it might be the last one he was going to enjoy. He had no birthday feast except some bits of ham-canvas—no luxury but this, and no substantials save the leather and oaken bucket-staves.

Speaking of the leather diet, one of the men

told me he was obliged to eat a pair of boots which were so old and rotten that they were full of holes; and then he smiled gently and said he didn't know, though, but what the holes tasted about as good as the balance of the boot. This man was very feeble, and after saying this he went to bed.

AFTERWORD

Richard Bucci

In early May 1867, this little volume began appearing in bookshops around New York. It measured just four and a half inches by not quite seven inches, and had, in the words of its author, a "gorgeous gold frog" pressed on its cloth cover. One thousand copies had been printed. *The Celebrated Jumping Frog of Calaveras County, And other Sketches** was Mark Twain's first book, and it was published by his friend Charles Henry Webb in New York. It collected twenty-seven sketches the author wrote while he was living in California and Nevada. The earliest piece, "Curing a Cold," dates from September 1863, and the last, "Concerning Chambermaids," was first printed in December 1866. The sketches had originally appeared in eight different daily and weekly newspapers. Foremost among these were the *Virginia City (Nevada) Territorial Enterprise*, the principal school where Mark Twain served his apprenticeship as a writer, and the *Californian*, a San Francisco literary weekly of some importance, which had been edited by Webb and Bret Harte. The reputation Mark Twain gained in the pages of these two journals brought him other assignments, and five of the sketches in this book, including the title sketch, were printed in New York first, as contributions to the *Saturday Press* and the *Weekly Review*, two small but respected literary papers.

Mark Twain ended his western days in December 1866. He sailed from San Francisco southward in an "Opposition steamer," crossed over the isth-

*In this essay, spelling and punctuation in nineteenth-century titles and quotations have not been made regular or modern.

mus at Nicaragua, and arrived in New York in January 1867. Within days he was reunited with Charles Henry Webb, who had preceded him there by eight months. Mark Twain was full of plans. He had had success lecturing in the West about Hawaii, whereto he had recently traveled as correspondent for the *Sacramento Union*. He wanted to lecture in New York, too, and to publish his Hawaii letters in a book. He eventually arranged three lectures (including one at Cooper Union), but was forced to scrap the Hawaii book when he could find no publisher for it. Instead, at Webb's suggestion, he began to fashion another book, by selecting and revising sketches from his scrapbook of California writings. He took a rudimentary manuscript to George Carleton, a likely publisher, but was refused by him. Mark Twain later recalled that Webb then stepped in. He "bravely said that not all the Carletons in the universe should defeat that book; he would publish it himself on a ten per cent. royalty. And so he did."[1]

When *Jumping Frog* came out, Mark Twain described Webb's role in its creation in a letter to a California newspaper.

> Webb ("Inigo") has fixed up a volume of my sketches, and he and the American News Company will publish it on Thursday, the 25th of the present month. He has gotten it up in elegant style, and has done everything to suit his own taste, which is excellent. I have made no suggestions. He calls it "THE CELEBRATED JUMPING FROG, AND OTHER SKETCHES, by 'Mark Twain.' Edited by C. H. Webb." Its price is $1 50 a copy. It will have a truly gorgeous gold frog on the back of it, and that frog alone will be worth the money. . . . Mail your orders either to C. H. Webb or the American News Company, New York.[2]

The author was being overly generous, however, and his gratitude was not altogether genuine. It has been subsequently proven, for example, that Mark Twain did indeed make "suggestions," helping Webb to select and revise the sketches for the book.[3] By proclaiming Webb the sole compiler and editor, he hoped to insulate himself from possible detractors among his new eastern audience. It remains true that without Charles Henry Webb there would have been no Jumping Frog book in 1867. However, at times Mark Twain thought that no book might not have been a bad thing.

Not that the young author was handled roughly by the critics. On the contrary, with few exceptions *Jumping Frog* was received with favor, and often with considerable enthusiasm.[4] The *New York Times* was perhaps most reserved in its praise.

> There is a great deal of quaint humor and much pithy wisdom in his writings, and their own merit, as well as the attractive style in which they are produced, must secure them a popularity which will bring its own profit. ("New Publications," 1 May 1867, 2)

The reaction in the New York *Citizen*, a small literary weekly, was more typical.

> Mark Twain is a genuine humorist. He does not depend on misspelling, or on punning, for the comic effect of his sketches, but on legitimate wit and humor. Perhaps his chief characteristic is his habit of bringing two utterly incongruous things in close juxtaposition. He imitates no one, but his humor is thoroughly his own. ("Volumes Worth Reading, and Volumes to Be Avoided," 4 May 1867, 4)

Wilkes Spirit of the Times, a New York magazine with a purported circulation of twenty thousand, favorably compared Mark Twain's lectures to Artemus Ward's, and remarked that the Jumping Frog book was "by long odds the funniest production out since 'Phenixiana,' and places its author in the color-guard of American humorists."[5] The New York *Dispatch*, which also claimed twenty thousand readers, believed Mark Twain was ahead of even such funny men as Artemus Ward or John Phoenix (George Horatio Derby, late author of *Phœnixiana*).

> It is not often that a genuinely humorous book is presented to the public — a book that one can read from beginning to end without experiencing a sense of weariness, not to say disgust. Occasionally a case of the kind does

> occur, but the instances are so rare that they deserve special comment. And for this reason we propose to devote a little space to the book which forms the subject of this notice. Of the great army of humorists, we have always placed Mark Twain at the head, and it is, we believe, universally conceded that his quiet wit, forcible hits, and unwavering pleasantry, combined with a certain gravity of expression peculiar to himself, are points not to be found in other funny writers of the day, and are as admirable as they are scarce. In the book now in hand we have a collection of sketches, many of them originally written for newspaper publication; and we may here state that the contents of this little work lose none of their gist and sparkle, and are none the less valuable from having been previously read, for what Mark Twain writes, will bear many perusals. ("New Publications," 5 May 1867, 7)

However, neither the considerable critical acclaim awarded to the book, nor even the accolades and insightful recognition directed at its author, could entirely mitigate Mark Twain's uneasy attitude toward his creation. Certainly the low American sales — only four thousand or so copies were bound in the three years it was in print — would cause him distress.[6] But long before its modest performance in the marketplace was apparent, Mark Twain expressed dissatisfaction with aspects of *Jumping Frog*. On the day of publication, he carped to Bret Harte that it was "full of damnable errors of grammar, & deadly inconsistencies of spelling in the Frog sketch."[7] Later on, when he especially feared the disapproving judgment of Olivia Langdon, his future wife, he denounced the book categorically.

> *Don't* read a word in that Jumping Frog book, Livy — *don't*. I hate to hear that infamous volume mentioned. I would be glad to know that every copy of it was burned, & gone forever. I'll never write another like it.[8]

Mark Twain's ambivalence about his literature surpassed mere anxiety an author might have over possible negative critical reaction, or the prospect of poor sales. In fact, he was not struggling with fear of failure, but rather, with fear of a certain kind of success. He was not sure that he wanted to be a popular humorist. In a remarkable letter to his brother, written in San Francisco in 1865, the future giant of American prose expressed his dilemma with power and economy.

> I *have* had a "call" to literature, of a low order — *i.e.* humorous. It is nothing to be proud of, but it is my strongest suit, & if I were to listen to that maxim of stern *duty* which says that to do right you **must** multiply the one or the two or the three talents which the Almighty entrusts to your keeping, I would long ago have ceased to meddle with things for which I was by nature unfitted & turned my attention to seriously scribbling to excite the **laughter** of God's creatures. Poor, pitiful business![9]

This moment of self-revelation came upon him just as he was writing down a tale he had heard in a dreary California mining camp, about how a local betting man received his comeuppance at the hands of a visiting stranger. The author sent his sketch, which he called "Jim Smiley and His Jumping Frog," to Artemus Ward (Charles Farrar Browne) in New York. Browne, then America's most popular humorist, intended to include it in his second book, *Artemus Ward: His Travels*. When the sketch arrived late, George Carleton, who was Browne's publisher, passed it on to Browne's friend and mentor, the critic Henry Clapp Jr. Clapp published "Jim Smiley and His Jumping Frog" in the New York *Saturday Press* on 18 November 1865. The New York correspondent of a California newspaper witnessed the ensuing success.

> Mark Twain's story . . . has set all New York in a roar, and he may be said to have made his mark. I have been asked fifty times about it and its author, and the papers are copying it far and near. It is voted the best thing of the day.[10]

Thus did Mark Twain gain a modest but important notoriety in the East, but from what he privately described as "a villainous backwoods sketch."[11]

It is certainly understandable that Mark Twain, then a "western writer" and a "humorist," should have been leery of entering the larger literary scene. In his imagination at least, either designation might cause sophisticated eastern critics to discount his work. But why should the author have retained a self-deprecatory attitude toward his talents *after* his book had so conquered the New York critics? Simply put, the answer is that while Mark Twain welcomed the praise he received in New York, literary taste in America in the 1860s, including his own, was still being made in Boston. The attitude that prevailed at the time was neatly suggested in a memoir by Mark Twain's future friend W. D. Howells. Speaking of the year 1866, Howells wrote:

> . . . there was in those days very little good writing done beyond the borders of New England. . . . The literary theories we accepted were New England theories, the criticism we valued was New England criticism, or, more strictly speaking, Boston theories, Boston criticism.[12]

Boston's aesthetic reign was established by men like Emerson, Hawthorne, Holmes, Longfellow, and Whittier, who forced a dichotomy in American literature. Their philosophic outlook, artistic judgment, and literary product not only formed the basis of the superior category of the dichotomy, but also defined the opposite, inferior category, by what they excluded. What was not "high" and "serious" became "low" and "humorous." When Henry Nash Smith explored the structure of the split some years ago, he regarded it as having developed of the general literate populace's competing desires to adopt and also to reject conventional moral and ethical assumptions. Positive, conventional moral philosophy, said Smith, was found in polite, or "genteel," literature — a word he borrowed from philosopher George Santayana. The negative, deflating philosophic position, which undermined conventional morality, found its expression in humorous literature. While some critics had simply regarded genteel literature as "eastern," and humorous writing as "western," Smith downgraded the usefulness of these designations, noting that if "geography has any meaning in this connection, it lies in the fact that spokesmen for the conventional American culture attempted to maintain a connection with Europe."[13] Smith's point is that practitioners of "low," vernacular humor could be found in New England as well as in California or the Southwest, just as conventional morality was proclaimed by writers on both shores of the American continent.

What must not be lost in Smith's consideration, however, is that the makers of the cultural code did come from New England. Their aesthetic judgments were the outcome, at least in part, of that region's puritanical heritage. The European link which the New Englanders sought to maintain was with old England, their ancestral home.

Because Boston's puritanical, Anglophilic values dominated the cultural terrain in America for many years, Mark Twain, like many others outside Boston, internalized these values, adopting them to one or another extent. His identification with a pattern of taste that was essentially alien to his nature and at odds with his talents led him (and Webb) to select conservatively from his western writings when compiling the Jumping Frog book, and also to fret over how he might come to be regarded once it was published. The author, however, came to national recognition in New York and not in Boston.

In New York before the Civil War, the aesthetic dichotomy was beginning to break up; the strict categorical divisions were occasionally abrogated. New York had never been a seat of gentility, having been established by Dutch traders rather than English sectarians. Just as puritanical morality had endured for two centuries in New England, the financial and mercantile spirit remained alive in New York long after the Dutch had been driven out of Manhattan. The city flourished in the nineteenth century, and at the time of the Civil War its population approached one million, three times that of Boston. As the country moved into its age of massive industrial expansion as a result of the war, aesthetic determinations made in Boston began to lose their universality.

Mark Twain himself would, of course, help to redefine what Americans regarded as serious literature, but he could not have known this in 1867. Similarly, he seems to have been unaware that the reason his first book was welcomed in the city of its publication was that literary society there had outgrown the Boston aesthetic paradigm, and looked to places other than London for inspiration. In New York the negative deflators of conventional morality held a high place on the cultural scene, and they too claimed a link to Europe, but not to England. Beginning in the 1850s, an influential circle of New York humorists, critics, poets, and thespians claimed as their own the more sensual, rebellious intellectualism of Bohemian Paris. And the man most responsible for bringing Bohemianism to America was none other than Henry Clapp Jr., publisher of the "Jumping Frog" sketch.

A BOHEMIA ON BROADWAY

The story of nineteenth-century American Bohemianism is necessarily informal, and even arcane. "Bohemianism" is itself a difficult word with a long history, but it describes, or was made to describe, the way of life among the intellectual *petits bourgeois déclassés* of Paris in the 1830s and 1840s. The period coincided with the reign of Louis Philippe, the "bourgeois king," whose rule encouraged the ascendancy of French capitalism. The Bohemians came from families of artisans and shopkeepers, whose livelihood was squeezed by the rising wealth of the bourgeoisie. They therefore stood in opposition to the new materialism, and proclaimed a revolt — sometimes coherent, sometimes not — in art and politics. The 1848 revolution and its aftermath of triumphant reaction brought an end to the Bohemian revolt, yet even in the flattened political landscape of the Second Empire, the image, the style of Bohemianism endured.

Clapp (1814–1875) traveled to Paris in the 1850s and absorbed the Bohemian attitude in its somewhat truncated form, and he later transplanted this attitude to New York City. Clapp was a Nantucketer by birth and the son of a minister. In his early days he had campaigned for temperance and abolition; when he lived in Paris he began to enjoy the sensual pleasures. He frequented cafés and dance halls, gave up the notion of temperance, and started to drink freely. He did not abandon his political convictions, however, but rather deepened them. He studied the works of Charles Fourier (which he later helped Albert Brisbane introduce to America), and he absorbed the irreverent criticisms of the Parisian feuilletonists. When he returned to America, he made New York his home, and established his headquarters at a long table in Pfaff's, a downstairs German restaurant at 653 Broadway. Beginning in the late 1850s, New York's creative talent gathered about him for nightly discussion. At the core of his group were the critics Thomas B.

Aldrich, Ned Wilkins, Edward House, Charles B. Seymour, and William Winter, the poets George Arnold, Fitz-James O'Brien, and Edmund Clarence Stedman, and illustrators Sol Eytinge and a very young Thomas Nast. Clapp encouraged women to brave his subterranean vault also, and the poet Ada Clare (Jane McElhinney), who had just returned from Paris herself, became one of his favorite companions. Actress Adah Isaacs Menken came downstairs too, sometimes with her third husband, *Sunday Mercury* editor Robert Henry Newell ("Orpheus C. Kerr"), and sometimes not. Fitz Hugh Ludlow, author of the sensational book *The Hasheesh Eater*, and none other than Artemus Ward himself, when he was editing *Vanity Fair* (a favorite repository of Bohemian literature), were also at Clapp's table. To all this genius, Henry Clapp was "King of Bohemia."

He was no mere carouser; he took literary life more seriously than he got credit for. In 1858 he founded the *Saturday Press*, and filled its pages with stories, essays, and poems by the best writers in New York. One of his contributors was Walt Whitman. At a time when *Leaves of Grass* had twice issued and had been twice ignored, its author found a place in Clapp's vault, and a friend and champion in Bohemia's editor. From its inception, the *Saturday Press* was with Whitman and against his many detractors. As an old man, Whitman told Horace Traubel, his chronicler,

> Henry Clapp stepped out from the crowd of hooters—was my friend: a much needed ally at that time (having a paper of his own) when almost the whole press of America when it mentioned me at all treated me with derision or worse. If you ever write anything about me in which it may be properly alluded to I hope you will say good things about Henry Clapp: indeed, I charge you to say them. To ignore him, or to say what should not be said about him, would be more unjust to me than to him.[14]

In his memoir, W. D. Howells, who had been one of the earliest contributors to Clapp's paper, described its influence.

> ... I had written for the *Saturday Press* myself, and had my pride in it, and perhaps it was because that paper really embodied the new literary life of

> the city. It was clever, and full of the wit that tries its teeth upon everything. It attacked all literary shams but its own, and it made itself felt and feared. The young writers throughout the country were ambitious to be seen in it, and they gave their best to it; they gave literally, for the *Saturday Press* never paid in anything but hopes of paying, vaguer even than promises. It is not too much to say that it was very nearly as well for one to be accepted by the *Press* as to be accepted by the *Atlantic*, and for the time there was no other literary comparison. To be in it was to be in the company of Fitz James O'Brien, Fitzhugh Ludlow, Mr. Aldrich, Mr. Stedman, and whoever else was liveliest in prose or loveliest in verse at that day in New York.[15]

Howells, however, was a would-be Bostonian in those days, and he soon fled the *Saturday Press*, because he "had found there a bitterness against Boston as great as the bitterness against respectability."[16]

BOHEMIA GOES WEST

The Bohemian moment in New York was suspended by the war, as graver concerns than literature gripped the Pfaffians. Some enlisted in the Union army, and some found work as war correspondents for New York newspapers. Eventually, a good many sought temporary refuge from the blood-soaked East, and so they toured the Far West, beginning in 1863. Among the first of Clapp's circle to arrive in San Francisco was the young war correspondent for the *New York Times*, Charles Henry Webb. Webb established himself first as city editor of the *San Francisco Evening Bulletin*, and then as Bret Harte's colleague at the *Golden Era*, at the time the only literary paper of significance in the West. Webb and Harte were not at all alike, but they clung to each other from the first. Harte was a local talent whose art stood out from the imitative effusions of other writers around him. Since 1857 he had received high praise for his dozens of contributions to the *Golden Era*, but he was nonetheless aware that his San Francisco milieu was yet a cultural backwater. Webb brought a whiff of true metropolitanism to the far shore, and Harte welcomed his arrival.

Additional Pfaffians began to concentrate in the West in late 1863, and quickly transformed its cultural life. The most celebrated visitor was Artemus Ward. His autumn lectures in California and Nevada were easily the most important entertainments ever given there. Noteworthy too was the arrival of Adah Isaacs Menken, who came with Newell, her hapless but talented husband, in tow. "The Menken" gave a number of dramatic performances in Virginia City and San Francisco, including a daring display in a melodrama based on Byron's *Mazeppa*, wherein she thrilled the mostly male audiences by simulating nudity in a flesh-colored body suit. Clapp's "Queen of Bohemia," Ada Clare, also came west. She was accompanied by her beloved young son, whom she had borne in Paris — the issue of an illegitimate union with the notoriously frisky pianist and composer Louis Moreau Gottschalk (who also turned up). The "Hasheesh Eater," Fitz Hugh Ludlow, arrived, along with the noted painter Albert Bierstadt. The editor of the *Golden Era*, Joseph Lawrence, knew a good thing when he saw it, and he threw open the columns of his journal to the eastern visitors. The locally produced sentimental verse and serialized romances were soon competing for space in the *Golden Era* not only with the contributions of Bret Harte and Charles Henry Webb, but also with the essays of Ludlow and Ada Clare, the satires of Newell (writing as "Orpheus C. Kerr"), Menken's poetry, and above all, the sketches of Artemus Ward. Thus San Francisco's first significant literary awakening was almost entirely imported from New York, an accident caused by the Civil War.

Few western writers could withstand comparison with the literary transients. In fact, aside from Bret Harte's contributions, only the work of William Wright stood out. Wright lived not in San Francisco but in Virginia City, Nevada, and he wrote under the name "Dan De Quille." He began contributing articles to the *Golden Era* in 1862, and in May of that year joined the staff of the *Virginia City Territorial Enterprise*, a position he would hold for thirty years to come. Like Bret Harte, De Quille wrote of the rugged life in the mining regions, but unlike his San Francisco rival, he was at home in this milieu. Intentionally, Harte betrayed in his work an aloof attitude toward his subjects, always signaling to readers that he, Harte, was not of the world he was describing. Not so De Quille. He was the genuine article, entirely comfortable on the dusty new frontier. His language was plain and vigorous. He sketched his characters at close hand, and exaggerated their traits to comic effect with evident sympathy. De Quille's "sage brush" journalism became popular in San Francisco precisely because it was true to life: nearly every clerk and waiter in the city had only recently quit the mines, and recognized in his stories characters from their own experience.

De Quille's success at the *Golden Era* certainly helped the career of the man who would soon eclipse him, Mark Twain. By a few months, he was De Quille's junior on the staff of the *Enterprise*. In September 1863, Mark Twain came down to San Francisco and stayed for a month. He took up with the visiting Bohemians, who recognized in him — despite his small literary output to that time — an outstanding talent and a kindred spirit. He met Joseph Lawrence and agreed to contribute to the *Golden Era*. He met Fitz Hugh Ludlow, and when Ludlow left San Francisco in November 1863, he wrote that the newcomer "is in a school by himself. He imitates nobody."[17] Most importantly, he so impressed Charles Henry Webb that when Webb began contemplating a literary journal of his own, he thought that Harte and Mark Twain would be the likely contributors.[18] Finally, back in Virginia City, he won over Artemus Ward, who had come there to lecture in December. Mark Twain quickly became his favorite companion, and by the time Artemus departed the two men were fast friends.

These early perceptions of Mark Twain, articulated by some of the leading lights of subversive culture, were remarkably acute. His genius was not yet revealed, but an aspect of his character already set him apart from possible western rivals. On the *Enterprise*, he wrote as well as De Quille; his prose possessed the same vigor, and he demonstrated the same easy familiarity with the crude new world he described. De Quille, however, was perhaps too close to that world, and could never transfer his talents to other subjects, other worlds. Mark Twain, on the other hand, had sought new vistas for his entire brief life. Raised in a tiny Mississippi river town where he learned the printing trade, he ran away to the East at the age of seventeen. For six months, he drank in the culture and sights of New York, Philadelphia, and Washington, working the while as a "jour" printer on metropolitan publications. When he returned to Missouri he became a river pilot, no mean feat and the dream of every boy in the region. When his brother was appointed secretary of the Nevada Territory by President Lincoln, he went west as his assistant, and then abandoned politics for gold and silver mining. All along the way, he continued to demonstrate considerable literary promise, composing local reports, sketches, and travel letters, mainly for newspapers back home. Now, with equal enthusiasm, the author plunged into the den of the eastern Bohemians, and conquered it in a few weeks. In doing so, he not only left De Quille behind, but also showed his superiority over Bret Harte.

Harte was acknowledged as the West's most important writer, a status he would enjoy for several more years. But whereas Mark Twain freely conceded that he had much to learn from Harte, Harte did not think he had much to learn from anybody. He enjoyed his role as the biggest fish in a little pond, and held himself aloof from literary colleagues. His status had also brought him material comfort: patrons in San Francisco society secured for him the position of secretary to the superintendent of the Mint. Thus, unlike his colleagues, Harte did not have to write to eat. Satisfied with his celebrity and prosperity, he was quick to rest on his laurels and even adopted, at the age of twenty-seven, the manner of a successful bourgeois. Harte's snobbery worked in much the same way as De Quille's parochialism; it cramped his talent and finally arrested it altogether.

Mark Twain was mainly free of these encumbrances, and he proved this in San Francisco, where he came to live in 1864. He put in a solid two years of writing there, and when he left in 1866, his fame rivaled Bret Harte's. He began as a city reporter for the *San Francisco Morning Call*, whose offices were upstairs from the San Francisco Mint. The job offered him little more than a salary, and he quit after a few months. But he did become acquainted with Bret Harte during this time, and the two men often swapped stories at Harte's desk in the Mint office.

Soon after leaving the *Call*, Mark Twain began to contribute to the *Californian*. Webb had launched his weekly paper just as Mark Twain was arriving in San Francisco, but then the author's grueling routine at the *Call* kept him from contributing. Webb and Harte began to conceive of an alternative to the *Golden Era* during and just after the visits of the eastern Bohemians. The brief presence of Webb's former cronies convinced both men that San Francisco needed a more critical and selective literary periodical. Webb copied the design of the *Californian* from the *Round Table*, a recently founded New York paper, and he and Harte supplied the bulk of the original articles. A few other local writers—notably the poet Charles Warren Stoddard—were encouraged to contribute, but Webb filled most of the remaining column space with reprintings from his favorite New York, London, and Parisian journals.

Mark Twain's involvement with Webb's new project is a milestone in his literary career. He would contribute twenty-five times to the *Californian* in two years, and taken as a whole, this work is more purely literary than any other body of writing he did in the West. It is no surprise, then, that when they came to make the Jumping Frog book, author and editor selected from the *Californian* more frequently than from any other source. Some of the most effective and enduring selections—"Answers to Correspondents," "An Item which the Editor Himself could not Understand," and "The Launch of the Steamer Capital"—appeared in the *Californian* first.[19]

Mark Twain interrupted his work for Webb several times before he left the West for good in December 1866. The first hiatus was in December 1864, when he began a three-month sojourn in the California mining region of Tuolumne and Calaveras counties. He published nothing during the rainy winter, mainly huddling inside remote cabins with a variety of rugged fortune-seekers, and listening to them tell stories to kill time. The author would make use of a number of these tales in his subsequent work, but none so successfully as "Jumping Frog," which we regard today as one of the most effective pieces of humorous writing in American literature. When Mark Twain returned to San Francisco, he resumed contributing to the *Californian* but had no other regular work. For more than six months he struggled with poverty and considered his future course. The moment of resolution, already described, came to him suddenly, as he wrote down the "Jumping Frog" tale and then informed his brother that he had reluctantly accepted his literary destiny. Nevertheless, he was still unsure what his future held, and in that same momentous letter to his brother, he threatened suicide if he could not write his way out of poverty.[20]

Write he did — mainly as San Francisco correspondent for the *Virginia City Territorial Enterprise*. The editor of the *Enterprise*, Joseph Goodman, was a subtly intellectual man. Under his guidance the *Enterprise* had become arguably the best newspaper in the West, despite its remote and dusty place of publication. And Goodman had always recognized that his young reporter was potentially a great author. He offered Mark Twain one hundred dollars a month for a *daily* letter about city life. It was a formidable assignment, but Mark Twain accepted, and for the next five months wrote like he never had before. It is a small tragedy that most of this work does not survive: the files of the *Enterprise* for this period were destroyed by fire. What has come down to us — about a quarter of the total — is very good. If the author's work for the *Californian* is his most self-consciously literary work of this period, his much larger accomplishment for the *Enterprise* is the most dazzlingly rich and creative. Unlike the editor of the *Call*, Goodman was able to give his man a free rein, and Mark Twain gave back two thousand words a day of muckraking reports of government corruption, humorous anecdotes, dramatic criticisms, and Mississippi reminiscences. Unfortunately, neither Mark Twain nor Webb could find a way to include much of this work in the Jumping Frog book. They seem to have regarded the material as mostly too topical and not literary enough for an anthology. Nevertheless, they did make five selections, most prominently "Among the Spirits" and "Remarkable Instances of Presence of Mind."[21]

Mark Twain took one more important job before he left the West: in March of 1866 he became the Hawaii correspondent for the *Sacramento Union*. From twenty-five long letters the author wrote to the *Union*, Bret Harte pulled out three brief sketches and reprinted them in the *Californian*. All three — "Honored as a Curiosity in Honolulu," "The Steed 'Oahu,'" and "Short and Singular Rations" — made their way into this book. The author wrote most of the remaining selections — "A Complaint about Correspondents," "An Inquiry about Insurances," "Concerning Chambermaids," and the peculiarly Gothic-style tale "A Strange Dream" — for Henry Clapp's *Saturday Press* or the *Weekly Review*, with which Clapp was also associated. All four tales had been solicited by Bohemia after the success of "Jim Smiley and His Jumping Frog."[22]

The western sketches came east, in Webb's copies of the *Californian*, and in Mark Twain's scrapbook. In New York, the two men made the happy discovery that the Bohemian journals had been reprinting Mark Twain's California writings all along, mainly at Clapp's instigation. It is no surprise that the author and his editor decided to include in *Jumping Frog* almost all these reprintings, as well as most of the original writing Mark Twain had done for Clapp's journals. Ironically, the world that had brought the sketches public recognition was already passing from being. Bohemian New York did not quite survive the war. Henry Clapp's influence diminished as his talented protégés grew up, married, and sought stability and status. Attendance fell off at his nocturnal philosophizing sessions in the dingy basement on Broadway. The *Saturday Press* failed, and, sadly, the "King of Bohemia" finished out his days as a poor and forgotten person. For Mark Twain, Bohemia could never have been more than a brief way station in any case. He was in New York for no more than two months when he began making arrangements to move on; in June of 1867 he would sail for Europe and the Holy Land as traveling correspondent for the *San Francisco Alta California*. The genius that Clapp had recognized in him, like the genius in Whitman, was as expansive and universal as the art of the Bohemians was narrow and exclusive. For no matter how subversive Clapp and company stood in relation to the Boston literary establishment, as Bohemians they remained a small group of initiates whose creations mainly appealed to one another. Mark Twain rebelled against everything that was obscure in art, and everywhere sought to deepen and broaden his audience. His cause was not merely to deflate and criticize, but to *create*, in a new and democratic artistic language. The Jumping Frog book is only a small moment on his path to remaking American fiction, but it is the beginning moment — reason enough to justify the book's reappearance now, after one hundred thirty years. Grand historical significance aside, not a few of the sketches in this book, created in a remote past, still sparkle with their original humor and insight into the human condition.

This essay could not have been written without the encouragement of Robert H. Hirst, general editor of the Mark Twain Project at the University of California, Berkeley. During my tenure at the project, I enjoyed the fruits of his unsurpassed scholarship in the field of Mark Twain's early writing. I also gained much insight from the expert work of my colleagues there, editors Dahlia Armon, Robert Browning, Victor Fischer, Michael Frank, Lin Salamo, Kenneth Sanderson, and Harriet Smith.

NOTES

1. Autobiographical dictation dated 21 May 1906, MS in the Mark Twain Papers, University of California, Berkeley; published in *Mark Twain in Eruption*, Bernard DeVoto, ed. (New York: Harper and Brothers, 1940), 145.

2. "Letter from 'Mark Twain,'" *San Francisco Alta California*, 2 June 1867, 1 (letter 15, dated "New York, April 19th, 1867"); published in *Mark Twain's Travels with Mr. Brown*, Franklin Walker and G. Ezra Dane, eds. (New York: Alfred A. Knopf, 1940), 157–58.

3. Robert H. Hirst has convincingly demonstrated that Mark Twain and Charles Henry Webb made separate editorial contributions to *Jumping Frog*; see his textual introduction in *Early Tales and Sketches, Volume 1 (1851–1864)*, edited by Edgar M. Branch and Robert H. Hirst. The Works of Mark Twain (Berkeley: University of California Press, 1979), 503–46.

4. The most obvious exception was William Cullen Bryant's paper, the *New York Evening Post*, which called the book a "mistake." After conceding that the "Jumping Frog" story itself was good, the *Post* critic remarked that the "remainder . . . with one or two exceptions, is dreary, and often worse. It is not suited to this longitude" ("Gossip about Publishers and Books," 8 May 1867, 2).

5. "Mark Twain's Lecture," *Wilkes Spirit of the Times*, 18 May 1867, p. 193. Circulation figures are taken from *Geo. P. Rowell and Co.'s American Newspaper Directory* (New York: George P. Rowell and Co., 1869), 72, 74.

6. Hirst uncovered a bindery record for *Jumping Frog* in the Mark Twain Papers at the University of California at Berkeley. The record reveals that Webb printed 4,076 copies from 1867 to 1870; see *Early Tales and Sketches, Volume 1 (1851–1864)*, 545 n. 43. (Hirst also discovered that British piracies of *Jumping Frog* did far better than the American original; see 546–55.) By contrast, Mark Twain's second book, *The Innocents Abroad*, sold nearly seventy thousand copies in its first year (Robert H. Hirst, "The Making of *The Innocents Abroad*: 1867–1872," Ph.D. diss., University of California, Berkeley, 1975, 317).

7. Samuel L. Clemens to Bret Harte, 1 May 1867, in *Mark Twain's Letters, Volume 2 (1867–1868)*, edited by Harriet Smith, Richard Bucci, and Lin Salamo (Berkeley: University of California Press, 1990), 39.

8. Samuel L. Clemens to Olivia Langdon, 31 December 1868, in *Mark Twain's Letters, Volume 2 (1867–1868)*, 369–70.

9. Samuel L. Clemens to Orion and Mary E. (Mollie) Clemens, 19 and 20 October 1865, in *Mark Twain's Letters, Volume 1 (1853–1866)*, edited by Edgar M. Branch, Michael B. Frank, Kenneth M. Sanderson, et. al. (Berkeley: University of California Press, 1988), 322–23.

10. "Podgers' Letter from New York," dated 10 December 1865, *San Francisco Alta California*, 10 January 1866, 1.

11. Samuel L. Clemens to Jane Lampton Clemens and Pamela A. Moffett, 20 January 1866, in *Mark Twain's Letters, Volume 1 (1853–1866)*, 327.

12. W. D. Howells, *Literary Friends and Acquaintance: A Personal Retrospect of American Authorship* (New York: Harper and Brothers, 1901), 115.

13. Henry Nash Smith, *Mark Twain: The Development of a Writer* (Cambridge: Belknap Press of Harvard University Press, 1962), 3–4.

14. Horace Traubel, *With Walt Whitman in Camden (March 28–July 14, 1888)* (Boston: Small, Maynard and Co., 1906), 236.

15. Howells, *Literary Friends and Acquaintance*, 70.

16. Howells, *Literary Friends and Acquaintance*, 71.

17. Fitz Hugh Ludlow, "P. P. C., A Good-Bye Article," *Golden Era* 11 (22 November 1863): 4.

18. [Charles Henry Webb], "Things," *Golden Era* 11 (8 November 1863): 8. See also the discussion in the introduction to *Early Tales and Sketches, Volume 3 (1865–1867)* edited by Richard Bucci, Edgar M. Branch, and Robert H. Hirst. The Works of Mark Twain (Berkeley: University of California Press, forthcoming).

19. The other selections from the *Californian* — making ten in all — are "Aurelia's Unfortunate Young Man," "The Story of the Bad Little Boy who Didn't Come to Grief," "Literature in the Dry Diggings," "Lucretia Smith's Soldier," "The Killing of Julius Caesar 'Localized,'" "A Touching Story of George Washington's Boyhood," and "Origin of Illustrious Men."

20. "If I do not get out of debt in 3 months, — " he wrote, "pistols or poison for one — exit *me*." *Mark Twain's Letters, Volume 1 (1853–1866)*, 324.

21. The other *Enterprise* selections are "Among the Fenians," "'After' Jenkins," and "Brief Biographical Sketch of George Washington." A sixth *Enterprise* sketch, "Information for the Million," dates from the author's tenure as local reporter in Virginia City.

22. Three additional sketches, "Curing a Cold," "A Page from a Californian Almanac," and "Advice for Good Little Girls," first appeared, respectively, in the *Golden Era*, the *San Francisco Dramatic Chronicle*, and the *Youths' Companion*, a San Francisco paper for juveniles.

FOR FURTHER READING

Richard Bucci

The standard collection of Mark Twain's short writings is an ongoing series produced by the Mark Twain Project and published by the University of California Press in Berkeley. Already in print are *Early Tales and Sketches, Volume 1 (1851–1864)* (1979) and *Early Tales and Sketches, Volume 2 (1864–1865)* (1981). Both books are edited by Edgar M. Branch and Robert H. Hirst, with the assistance of Harriet Smith. A third volume, covering the years 1865–1867, edited by Branch, Hirst, and Richard Bucci, is due out in 1997. These books contain the most reliable texts of Mark Twain's sketches, as well as valuable historical and textual annotation. The Mark Twain Project and the University of California Press has also published three additional books of primary material from the author's early life. They are *Mark Twain's Letters, Volume 1 (1853–1866)*, edited by Edgar M. Branch, Michael B. Frank, Kenneth M. Sanderson, et al. (1988), *Mark Twain's Letters, Volume 2 (1867–1868)*, edited by Harriet Smith, Richard Bucci, and Lin Salamo (1990), and *Mark Twain's Notebooks and Journals, Volume I (1855–1873)*, edited by Frederick Anderson, Michael B. Frank, and Kenneth M. Sanderson (1975). Other collections of Mark Twain's early writings are John Howell, ed., *Sketches of the Sixties by Bret Harte and Mark Twain* (San Francisco: John Howell, 1927), Bernard Taper, ed., *Mark Twain's San Francisco* (New York: McGraw-Hill, 1963), and Franklin Walker, ed., *The Washoe Giant in San Francisco, Being Heretofore Uncollected Sketches by Mark Twain* (San Francisco: George Fields, 1938).

The standard biographical source for all periods of Mark Twain's life remains Albert Bigelow Paine's *Mark Twain: A Biography*, 3 vols. (New York: Harper and Brothers, 1912). The starting point for any discussion of the evolution of Mark Twain's art must be Henry Nash Smith's *Mark Twain: The Development of a Writer* (Cambridge: Belknap Press of Harvard University Press, 1962). The standard critical monograph on the author's early work is Edgar M. Branch's *The Literary Apprenticeship of Mark Twain. With Selections from His Apprentice Writing* (Urbana: University of Illinois Press,

1950). Indispensable for a study of the literary scene from which Mark Twain emerged is Franklin Walker's *San Francisco's Literary Frontier* (Seattle: University of Washington Press, 1939, rev. ed., 1969). The author's relationship with Bret Harte is discussed in George Stewart's *Bret Harte: Argonaut and Exile* (Boston: Houghton Mifflin, Riverside Press, 1931). His relationship with Charles Henry Webb is treated in Robert H. Hirst's textual introduction to *Early Tales and Sketches* (see vol. 1, 501–46), and in George Stewart's article on Webb in *Dictionary of American Biography*, Allen Johnson and Dumas Malone, eds., 20 vols. (New York: Charles Scribner's Sons, 1928–1936). For information on Henry Clapp and his Bohemian circle, no better source is available than Albert Parry's *Garrets and Pretenders: A History of Bohemianism in America* (New York: Covici-Friede, 1933).

A NOTE ON THE COVER

Beverly R. David & Ray Sapirstein

Although the volume is unillustrated, the gorgeous gold-stamped frog on the cover of Mark Twain's first book, *The Celebrated Jumping Frog of Calaveras County* (1867), merits brief mention. Appearing irregularly in different locations on the book cover, the frog is described by Merle Johnson, Twain's bibliographer, as having "a peculiar habit of jumping in various positions . . . but invariably fac[ing] the fore-edge."[1] With a nearly human body, the frog flies gracefully through space, wearing an expression of chagrin, unsure exactly why it is jumping, and somewhat dubious as to the specifics of how it will land.

Although Twain announced that he published the frog book only to "advertise myself," he did become involved in the editing and revision of the selections it contained.[2] Working under the pseudonym John Paul, C. H. Webb was the editor of the book and dealt with the tedious details of production and sales. When the slender volume appeared, Twain fumed that the text was filled with "damnable errors,"[3] but he was partial to the illustration on the cover declaring that "it would be well to publish the frog and leave the book out."[4] The famous jumping frog became a personal trademark and helped to spark in Twain what would become a sustained interest in book design and illustration.

NOTES

1. Merle Johnson, *A Bibliography of the Works of Mark Twain*, rev. ed. (New York: Harper and Brothers, 1935), p. 3.

2. Mark Twain, *Early Tales and Sketches, Volume 1 (1851–1864)*, Edgar M. Branch and Robert H. Hirst, eds. (Berkeley: University of California Press, 1979), pp. 545–46.

3. SLC to Bret Harte, May 1, 1867, in *Mark Twain's Letters, Volume 2 (1867–1868)*, Harriet Smith, Richard Bucci, and Lin Salamo, eds. (Berkeley: University of California Press, 1990), p. 39.

4. SLC to *San Francisco Alta California*, April 19, 1867, in *Mark Twain's Travels with Mr. Brown*, Franklin Walker and G. Ezra Dane, eds. (New York: Alfred A. Knopf, 1940), p. 289, n. 4.

A NOTE ON THE TEXT

Robert H. Hirst

This text of *The Celebrated Jumping Frog of Calaveras County, And other Sketches* is a photographic facsimile of a copy of the first American edition dated 1867 on the title page. The first edition was published on April 30 or May 1, 1867. The copy reproduced here is an example of Jacob Blanck's "second printing," or impression that is, unlike the very earliest copies, it lacks a page of advertisements inserted facing the title page, and it shows some minor type damage (21, 66, and 198) (*BAL* 3310). Still, this copy was among the first 2,000 printed, and it was probably bound sometime between May 20, 1867, and February 3, 1868. Only 4,076 copies were ever printed, because in December 1870 Mark Twain bought and destroyed the plates (*Early Tales and Sketches, Volume 1,* ed. Edgar M. Branch and Robert H. Hirst, The Works of Mark Twain, University of California Press, 1979, pp. 503–46). The original volume is in the collection of the Mark Twain House in Hartford, Connecticut (810/C625ce/1867/c. 3).

THE MARK TWAIN HOUSE

The Mark Twain House is a museum and research center dedicated to the study of Mark Twain, his works, and his times. The museum is located in the nineteen-room mansion in Hartford, Connecticut, built for and lived in by Samuel L. Clemens, his wife, and their three children, from 1874 to 1891. The Picturesque Gothic-style residence, with interior design by the firm of Louis Comfort Tiffany and Associated Artists, is one of the premier examples of domestic Victorian architecture in America. Clemens wrote *Adventures of Huckleberry Finn*, *The Adventures of Tom Sawyer*, *A Connecticut Yankee in King Arthur's Court*, *The Prince and the Pauper*, and *Life on the Mississippi* while living in Hartford.

The Mark Twain House is open year-round. In addition to tours of the house, the educational programs of the Mark Twain House include symposia, lectures, and teacher training seminars that focus on the contemporary relevance of Twain's legacy. Past programs have featured discussions of literary censorship with playwright Arthur Miller and writer William Styron; of the power of language with journalist Clarence Page, comedian Dick Gregory, and writer Gloria Naylor; and of the challenges of teaching *Adventures of Huckleberry Finn* amidst charges of racism.

CONTRIBUTORS

Roy Blount Jr. is the author of *About Three Bricks Shy—and the Load Filled Up* (1974), *Crackers* (1980), *One Fell Soup* (1982), *What Men Don't Tell Women* (1984), *Now, Where Were We?* (1988), *First Hubby* (1990), *Camels Are Easy, Comedy's Hard* (1991), and other books. He is a contributing editor of the *Atlantic Monthly* and the editor of *Roy Blount's Book of Southern Humor* (1994). He grew up in Decatur, Georgia, and attended Vanderbilt and Harvard. He lives in New York City and western Massachusetts.

Richard Bucci was for fourteen years an editor at the Mark Twain Project at the University of California, at Berkeley. His work includes *Mark Twain's Letters, Volume 2 (1867–1868)*, which he edited with Harriet Smith and Lin Salamo (University of California Press, 1990), and the forthcoming *Early Tales and Sketches, Volume 3 (1865–1867),* edited with Edgar M. Branch and Robert H. Hirst, also to be published by the University of California Press. Now a consulting editor with the project, he is also the author of *Richard Wright's "Native Son"* (1996) and "Is *Huckleberry Finn* a Racist Work?" (*Newsday*, 10 May 1985), and the editor of *The Prophet's Children: Travels on the American Left* by Tim Wohlforth (1994).

Beverly R. David is professor emerita of humanities and theater at Western Michigan University in Kalamazoo. She is currently working on volume 2 of *Mark Twain and His Illustrators*, and on a Mark Twain mystery entitled *Murder at the Matterhorn*. She has written a number of sections on illustration for the *Mark Twain Encyclopedia* and her *Mark Twain and His Illustrators, Volume 1 (1869–1875)* was published in 1989. Dr. David resides in Allegan, Michigan, in the summer and Green Valley, Arizona, in the winter.

Shelley Fisher Fishkin, professor of American Studies and English at the University of Texas at Austin, is the author of the award-winning books *Was Huck Black? Mark Twain and African-American Voices* (1993) and *From Fact to Fiction: Journalism and Imaginative Writing in America*

(1985). Her most recent book is *Lighting Out for the Territory: Reflections on Mark Twain and American Culture* (1996). She holds a Ph.D. in American Studies from Yale University, has lectured on Mark Twain in Belgium, England, France, Israel, Italy, Mexico, the Netherlands, and Turkey, as well as throughout the United States, and is president-elect of the Mark Twain Circle of America.

Robert H. Hirst is the General Editor of the Mark Twain Project at The Bancroft Library, University of California in Berkeley. Apart from that, he has no other known eccentricities.

Ray Sapirstein is a doctoral student in the American Civilization Program at the University of Texas at Austin. He curated the 1993 exhibition *Another Side of Huckleberry Finn: Mark Twain and Images of African Americans* at the Harry Ransom Humanities Research Center at the University of Texas at Austin. He is currently completing a dissertation on the photographic illustrations in several volumes of Paul Laurence Dunbar's poetry.

ACKNOWLEDGMENTS

There are a number of people without whom The Oxford Mark Twain would not have happened. I am indebted to Laura Brown, senior vice president and trade publisher, Oxford University Press, for suggesting that I edit an "Oxford Mark Twain," and for being so enthusiastic when I proposed that it take the present form. Her guidance and vision have informed the entire undertaking.

Crucial as well, from the earliest to the final stages, was the help of John Boyer, executive director of the Mark Twain House, who recognized the importance of the project and gave it his wholehearted support.

My father, Milton Fisher, believed in this project from the start and helped nurture it every step of the way, as did my stepmother, Carol Plaine Fisher. Their encouragement and support made it all possible. The memory of my mother, Renée B. Fisher, sustained me throughout.

I am enormously grateful to all the contributors to The Oxford Mark Twain for the effort they put into their essays, and for having been such fine, collegial collaborators. Each came through, just as I'd hoped, with fresh insights and lively prose. It was a privilege and a pleasure to work with them, and I value the friendships that we forged in the process.

In addition to writing his fine afterword, Louis J. Budd provided invaluable advice and support, even going so far as to read each of the essays for accuracy. All of us involved in this project are greatly in his debt. Both his knowledge of Mark Twain's work and his generosity as a colleague are legendary and unsurpassed.

Elizabeth Maguire's commitment to The Oxford Mark Twain during her time as senior editor at Oxford was exemplary. When the project proved to be more ambitious and complicated than any of us had expected, Liz helped make it not only manageable, but fun. Assistant editor Elda Rotor's wonderful help in coordinating all aspects of The Oxford Mark Twain, along with literature editor T. Susan Chang's enthusiastic involvement with the project in its final stages, helped bring it all to fruition.

I am extremely grateful to Joy Johannessen for her astute and sensitive copyediting, and for having been such a pleasure to work with. And I appreciate the conscientiousness and good humor with which Kathy Kuhtz Campbell heroically supervised all aspects of the set's production. Oxford president Edward Barry, vice president and editorial director Helen McInnis, marketing director Amy Roberts, publicity director Susan Rotermund, art director David Tran, trade editorial, design and production manager Adam Bohannon, trade advertising and promotion manager Woody Gilmartin, director of manufacturing Benjamin Lee, and the entire staff at Oxford were as supportive a team as any editor could desire.

The staff of the Mark Twain House provided superb assistance as well. I would like to thank Marianne Curling, curator, Debra Petke, education director, Beverly Zell, curator of photography, Britt Gustafson, assistant director of education, Beth Ann McPherson, assistant curator, and Pam Collins, administrative assistant, for all their generous help, and for allowing us to reproduce books and photographs from the Mark Twain House collection. One could not ask for more congenial or helpful partners in publishing.

G. Thomas Tanselle, vice president of the John Simon Guggenheim Memorial Foundation, and an expert on the history of the book, offered essential advice about how to create as responsible a facsimile edition as possible. I appreciate his very knowledgeable counsel.

I am deeply indebted to Robert H. Hirst, general editor of the Mark Twain Project at The Bancroft Library in Berkeley, for bringing his outstanding knowledge of Twain editions to bear on the selection of the books photographed for the facsimiles, for giving generous assistance all along the way, and for providing his meticulous notes on the text. The set is the richer for his advice. I would also like to express my gratitude to the Mark Twain Project, not only for making texts and photographs from their collection available to us, but also for nurturing Mark Twain studies with a steady infusion of matchless, important publications.

I would like to thank Jeffrey Kaimowitz, curator of the Watkinson Library at Trinity College, Hartford (where the Mark Twain House collection is kept), along with his colleagues Peter Knapp and Alesandra M. Schmidt, for having been instrumental in Robert Hirst's search for first editions that could be safely reproduced. Victor Fischer, Harriet Elinor Smith, and especially Kenneth M. Sanderson, associate editors with the Mark Twain Project, reviewed the note on the text in each volume with cheerful vigilance. Thanks are also due to Mark Twain Project associate editor Michael Frank and administrative assistant Brenda J. Bailey for their help at various stages.

I am grateful to Helen K. Copley for granting permission to publish photographs in the Mark Twain Collection of the James S. Copley Library in La Jolla, California, and to Carol Beales and Ron Vanderhye of the Copley Library for making my research trip to their institution so productive and enjoyable.

Several contributors — David Bradley, Louis J. Budd, Beverly R. David, Robert Hirst, Fred Kaplan, James S. Leonard, Toni Morrison, Lillian S. Robinson, Jeffrey Rubin-Dorsky, Ray Sapirstein, and David L. Smith — were particularly helpful in the early stages of the project, brainstorming about the cast of writers and scholars who could make it work. Others who participated in that process were John Boyer, James Cox, Robert Crunden, Joel Dinerstein, William Goetzmann, Calvin and Maria Johnson, Jim Magnuson, Arnold Rampersad, Siva Vaidhyanathan, Steve and Louise Weinberg, and Richard Yarborough.

Kevin Bochynski, famous among Twain scholars as an "angel" who is gifted at finding methods of making their research run more smoothly, was helpful in more ways than I can count. He did an outstanding job in his official capacity as production consultant to The Oxford Mark Twain, supervising the photography of the facsimiles. I am also grateful to him for having put me in touch via e-mail with Kent Rasmussen, author of the magisterial *Mark Twain A to Z*, who was tremendously helpful as the project proceeded, sharing insights on obscure illustrators and other points, and generously being "on call" for all sorts of unforeseen contingencies.

I am indebted to Siva Vaidhyanathan of the American Studies Program of the University of Texas at Austin for having been such a superb research assistant. It would be hard to imagine The Oxford Mark Twain without the benefit of his insights and energy. A fine scholar and writer in his own right, he was crucial to making this project happen.

Georgia Barnhill, the Andrew W. Mellon Curator of Graphic Arts at the American Antiquarian Society in Worcester, Massachusetts, Tom Staley, director of the Harry Ransom Humanities Research Center at the University of Texas at Austin, and Joan Grant, director of collection services at the Elmer Holmes Bobst Library of New York University, granted us access to their collections and assisted us in the reproduction of several volumes of The Oxford Mark Twain. I would also like to thank Kenneth Craven, Sally Leach, and Richard Oram of the Harry Ransom Humanities Research Center for their help in making HRC materials available, and Jay and John Crowley, of Jay's Publishers Services in Rockland, Massachusetts, for their efforts to photograph the books carefully and attentively.

I would like to express my gratitude for the grant I was awarded by the University Research Institute of the University of Texas at Austin to defray some of the costs of researching The Oxford Mark Twain. I am also grateful to American Studies director Robert Abzug and the University of Texas for the computer that facilitated my work on this project (and to UT systems analyst Steve Alemán, who tried his best to repair the damage when it crashed). Thanks also to American Studies administrative assistant Janice Bradley and graduate coordinator Melanie Livingston for their always generous and thoughtful help.

The Oxford Mark Twain would not have happened without the unstinting, wholehearted support of my husband, Jim Fishkin, who went way beyond the proverbial call of duty more times than I'm sure he cares to remember as he shared me unselfishly with that other man in my life, Mark Twain. I am also grateful to my family—to my sons Joey and Bobby, who cheered me on all along the way, as did Fannie Fishkin, David Fishkin, Gennie Gordon, Mildred Hope Witkin, and Leonard, Gillis, and Moss

Plaine—and to honorary family member Margaret Osborne, who did the same.

My greatest debt is to the man who set all this in motion. Only a figure as rich and complicated as Mark Twain could have sustained such energy and interest on the part of so many people for so long. Never boring, never dull, Mark Twain repays our attention again and again and again. It is a privilege to be able to honor his memory with The Oxford Mark Twain.

Shelley Fisher Fishkin
Austin, Texas
April 1996